CHEMISTRY OF NATURAL PRODUCTS

CHEMISTRY OF NATURAL PRODUCTS

K. ANAND SOLOMON
Senior Research Scientist
Sankar Foundation Research Institute
Vishakapatnam
Andhra Pradesh

MJP Publishers

Cataloguing-in-PublicationData

Anand Solomon, K. (1973-).
Chemistry of Natural Products / by K. Anand Solomon
Chennai : MJP Publishers, 2012
xii, 172 p.; 24 cm.
Includes References and Index.
ISBN 978-81-8094-129-0
1. Natural Products, Chemistry I. Title.
547 dc 22 ANA MJP 108

ISBN 978-81-8094-129-0

Printed and bound in India

MJP PUBLISHERS
New No. 5, Muthu Kalathy Street
Triplicane
Chennai 600 005

Publisher : J.C. Pillai
Managing Editor : C. Sajeesh Kumar
Project Editor : P. Parvath Radha
Acquisitions Editor : C. Janarthanan
Editorial Team : B. Ramalakshmi, V.R. Padma, R. Hemalatha, S. Jeevasruthi, M. Gnanasoundari, Lissy John, B. Annalakshmi
CIP Data : Prof. K. Hariharan, Librarian
RKM Vivekananda College, Chennai.

To

My beloved mother **A.P. Rajalakshmi**
who has been a source of inspiration, courage and love

Preface

Nature has been the best teacher and healer through centuries. The human race has evolved by observing Nature and healing through natural remedies. Science and Mind are two important tools in understanding Nature. Research in the area of natural products, i.e., products from Nature, be it of plant or animal origin, has been being carried out at a remarkable pace. Hence, isolation and characterization of natural products will help in understanding their mode of action with reference to their biological or pharmacological activity. Research in ayurveda, siddha or unani medicines, which involves natural products, scientifically may throw more light in curing illnesses that plague humankind. Particularly, a plant chemist or an ethnobotanist is involved in identification of the correct plant, an organic chemist extracts and isolates the components from the plant or a part of the plant and a scientist determines its biological or pharmacological activity. If results are promising, a group on Drug Discovery would start working on it. This book has been written with a feeling that it would help a bit in understanding the chemistry and importance of natural products.

First and foremost, I would like to thank my alma mater, Sri Sathya Sai University, Puttaparthi, which shaped my thinking process and career.

I have a special word of gratitude and love for Prof.S.S.Rajan, who was my mentor in research. From him I learnt the philosophy that knowledge is meant to be shared, and that simplicity is the hallmark of true education.

I thank my father Mr.R. Kamalakaran for all his support throughout my life . To Mrs.Janani, my beloved wife, I owe my love for her patience and perseverance.

I thank Dr.Gaddamanugu Gopi Krishna, my co-scientist for fruitful discussions and interactions.

To a soul-inspiring personality, Mr. Atmakuri Sankar Rao, who has established a super-speciality eye hospital that caters to the inflictions of the eye free of cost, I express my thanks for encouraging and supporting research in his foundation.

K. Anand Solomon

CONTENTS

1

INTRODUCTION

The term "natural products" generally connotes herbs, herbal concoctions, dietary supplements, traditional Chinese medicine, marine products, etc.[1] Natural products are generally either of prebiotic origin or originate from microbes, plants, or animal sources. As chemicals, natural products include such classes of compounds as terpenoids, polyketides, amino acids, peptides, proteins, carbohydrates, lipids, ribonucleic acid (RNA), deoxyribonucleic acid (DNA) and so forth.

(a)

(b)

Figure 1.1 Some natural products (a) Seeds of *Nux vomica*—a violent purgative (b) Root of *Rawvolfia serpentina*—an antihypertensive

But to be specific, natural products refers to chemicals known as secondary metabolites. These are generally produced by plants (Figure 1.1) and a few other organisms and are not essential for normal growth, development or reproduction of an organism, unlike proteins and nucleic acids. A limited number of species produce secondary metabolites. These compounds attract scientific interest because they usually exert an influence over other living organisms, often playing a part in directing interactions between plants, microorganisms, insects and animals.[2] Defence chemicals, anti-feedants, attractants and pheromones are all natural products and belong to this category.

For example, flavonoids are a group of naturally occurring polyphenolic compounds ubiquitously found in the plant kingdom.[3] They are widespread in vegetables, fruits, flowers, seeds, and grains.[4] The exact role of these secondary metabolites is still unclear, but it is known that flavonoids are important for the survival of a plant in its environment; they regulate plant growth, inhibit or kill many bacterial strains, inhibit major viral enzymes, and destroy some pathogenic protozoans.[5]

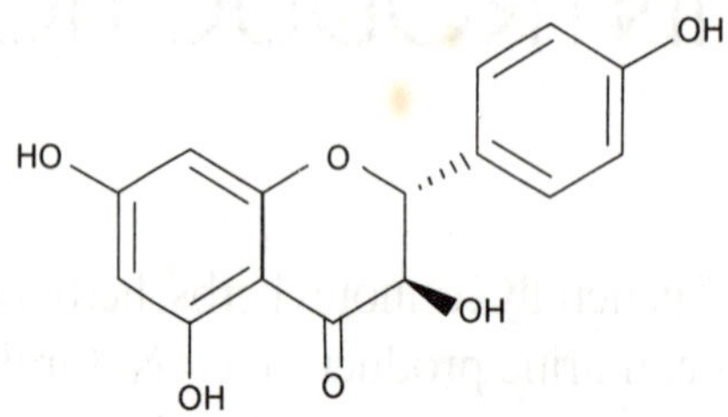

Taxifolin—a flavonoid derivative

Also, they act in plants as visual attractors, feeding repellents and photoreceptors, and protect against UV radiation.[6–8] Thus far, approximately 9000 different flavonoids have been identified and they form the largest group of naturally occurring polyphenols.[9]

Although plants are better known as a source of secondary metabolites, bacteria, fungi and many marine organisms (sponges, tunicates, corals, snails) too are very interesting sources. The function or importance of these compounds to the organism's development is usually of ecological nature as they are used as defence against predators (herbivores, pathogens, etc.), for interspecies competition, and to facilitate the reproductive processes.[10] Similarly, marine algae are also sessile and have no obvious means of defence. Defensive functions attributed to marine, secondary metabolites include predator detterence, anti-fouling, inhibition of overgrowth and ultraviolet radiation protection.[11]

It has been estimated that in certain marine ecosystems such as coral reefs or the deep sea floor, the biological diversity is higher than in tropical rainforests. Many marine organisms are soft-bodied and have a sedentary life style necessitating chemical means of defence. Therefore, they have evolved the ability to either synthesize toxic compounds or obtain them from marine microorganisms. These compounds help them deter predators, keep competitors at bay or paralyse their prey. This diversity is believed to give rise to an equally high diversity of secondary metabolites synthesized by the marine microfauna and microflora. For this reason, and because of the immense biological diversity in the sea as a whole, it is increasingly recognized that a huge number of natural products and novel chemical entities exist in the oceans, with biological activities that may be useful in the quest for finding drugs with greater efficacy and specificity for the treatment of many human diseases.

In general, natural products have several functional groups which are located into a precise three-dimensional position providing specific interactions with target molecules. It is often assumed that secondary metabolites in natural organisms have been optimized through evolution to exert their still-not-well-defined effects.

Secondary metabolites can be classified by their chemical structure or physical properties into one or more of the following groups:

- Lipids and derivatives
- Alkaloids
- Terpenoids
- Polypeptides
- Aliphatic, aromatic, and heteroaromatic organic acids
- Phenols
- Iridoids
- Steroids
- Carbohydrates
- Peptides
- Ethereal oils
- Resins
- Balsams

For millennia, medicinal remedies have exclusively been derived from the large pool of natural products. These natural cures were central to the therapy and prophylaxis of physical and mental diseases in the ancient agrarian and pre-industrialized cultures and societies. All the herbal-, mineral-, human- and animal-derived products, which supply the active ingredients of traditional medicines, are subsumed under the term *Materia Medica*, which were applied either as *simplicia* (i.e., consisting of one healing principle in a preparation matrix) or as composita (i.e., a combination of two or more medicinal drugs). The current accepted modern medicine or allopathy has gradually developed over the years by scientific and observational efforts of scientists. However, the basis of its development remains rooted in traditional medicine and therapies.

HISTORY OF NATURAL PRODUCTS

For more than thousand years natural products have been playing a pivotal role in the area of medicine, both for animals and human beings. The best-known Egyptian pharmaceutical record is the Ebers papyrus dating from 1500 BC, a document that describes around 700 drugs (mostly from plants). It includes formulas for gargles, snuffs, poultices, infusions, pills and ointments. The Chinese *Materia Medica* and Indian Ayurvedic systems have records dating back to 1100 BC and 1000 BC respectively.[10]

Plants have also been used in the production of stimulant beverages (e.g., tea, coffee, cocoa, and cola) and inebriants or intoxicants (e.g., wine, beer, kava) in many cultures since ancient times, and this trend continues till today. Tea (*Camellia sinensis* Kuntze) was first consumed in ancient China (the earliest reference is around CE 350), while coffee (*Coffeaarabica* L.) was initially cultivated in Yemen for commercial purposes in the 9th century.[10]

Numerous highly active chemical entities were isolated from their naturally occurring source. The first alkaloid isolated was morphine[11] from *Papaver somniferum* (opium) in 1804, known for its analgesic property. Quinine—an effective anti-malarial compound—was isolated from the bark of *Cinchona officinalis*[12] in 1820.

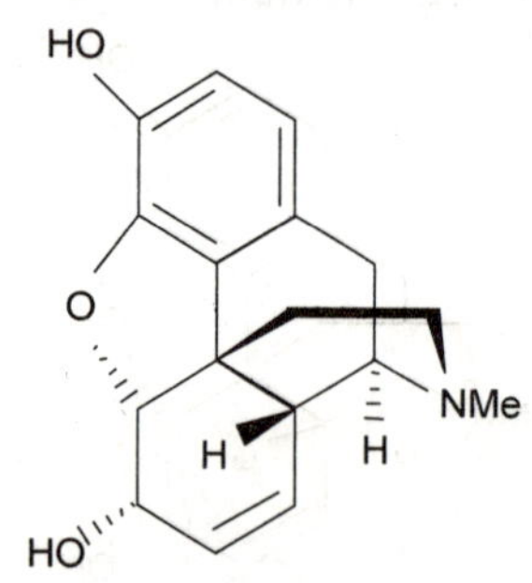

Morphine

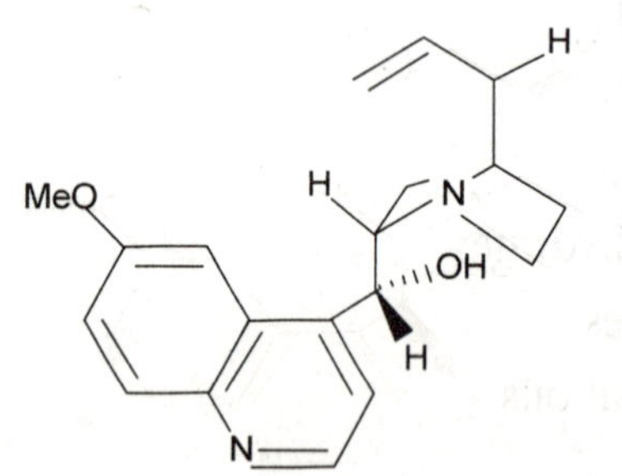

Quinine

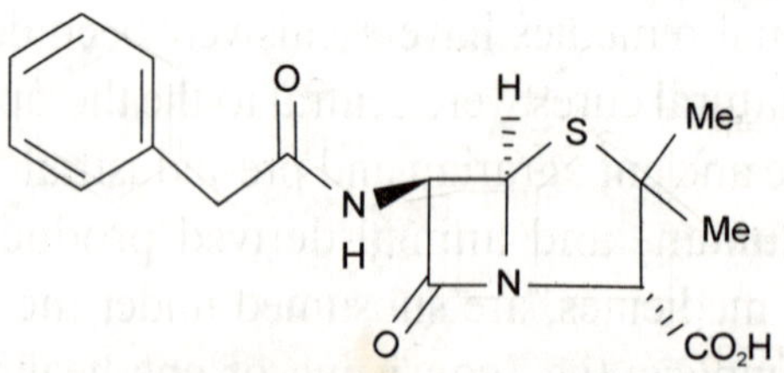

Penicillin G

The year 1929, saw the birth of pencillin isolated by Fleming from the fungus *Penicillium notatum.*[13] Vinblastine and vincristine isolated in 1954[14] from *Catharanthus roseus* were alkaloids developed as anti-cancer drugs by Eli Lily.

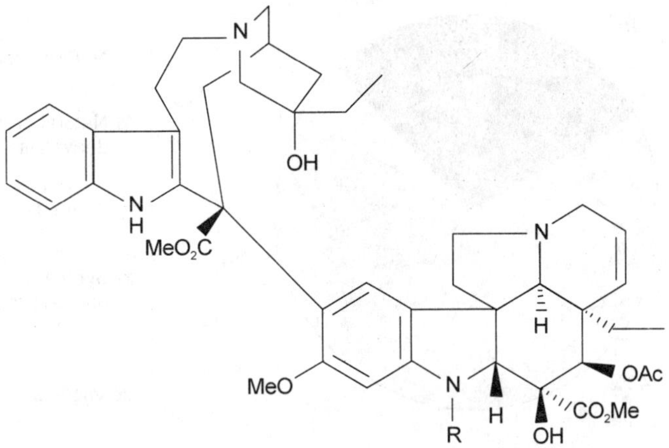

Vinblastine R= Me, Vincristine R=CHO

In 1969, paclitaxel was isolated from the Pacific yew tree (*Taxus brevifolia*) and is being marketed as an anti-cancer compound.[15]

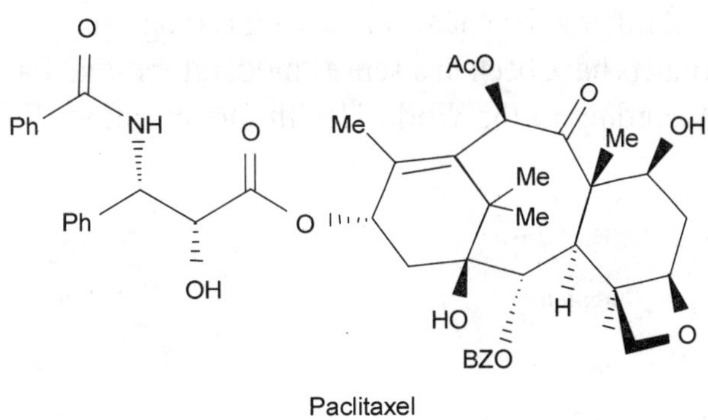

Paclitaxel

Apart from the above-mentioned compounds, the turn of 1800 marked the isolation of numerous alkaloids from plants (species in parentheses) used as drugs, namely, atropine (*Atropa belladonna*), caffeine (*Coffea arabica*), cocaine (*Erythroxylum coca*), ephedrine (*Ephedra* species), codeine (*Papaver somniferum*), pilocarpine (*Pilocarpus jaborandi* Holmes), physostigmine (*Physostigma venenosum*), salicin (*Salix* species), theobromine (*Theobroma cacao*), theophylline (*Camellia sinensis*), and (+)-tubocurarine (*Chondodendron tomentosum* Ruiz & Pav.).[16]

Figure 1.2 illustrates the role of natural products in pharmaceutical industry

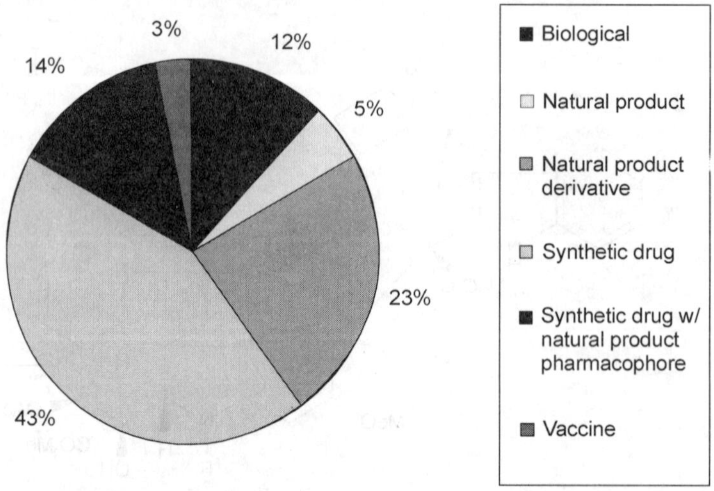

Figure 1.2 Pharmaceuticals containing natural products (1981–2002)[16]

SIGNIFICANCE OF NATURAL PRODUCTS

Chemical ecology and natural products chemistry are linked in a productive partnership (Figure 1.3) aimed at clarifying the chemical basis of ecological and behavioural interactions in nature. Natural products have been in a sense "moderators" that balance disease and health in human lifecycle. According to the World Health Organization (WHO), 80% of humanity relies primarily

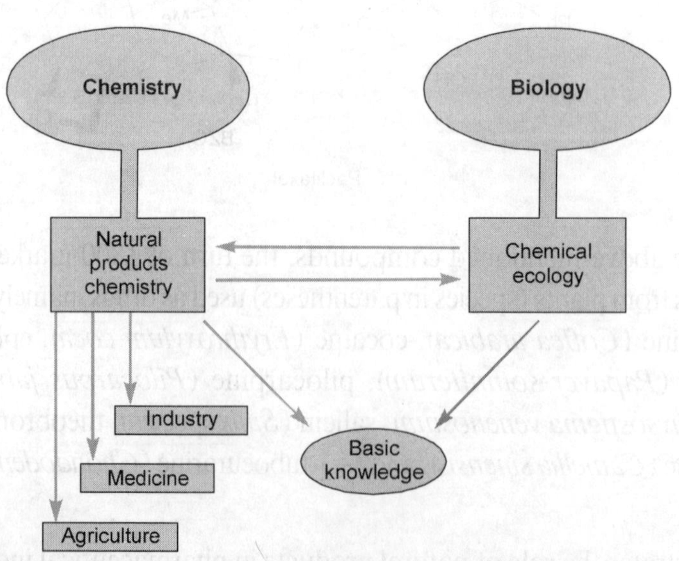

Figure 1.3 A depiction of the interrelation between chemistry and biology

on traditional medicines as sources for their primary health care.[17] Over hundred organic molecules considered to be important drugs have been derived from around hundred and fifty different plants. In the year 2000, of the twenty top-selling drugs in the market, which are not proteins, seven were either derived from natural products or developed from leads generated from natural products (Table 1.1) and many more are under process (Table 1.2).

Table 1.1 A few drugs of phytochemical origin

Drug	Year of introduction into market	Property	Company
Orlistat	1999	Antiobesity	Roche
Miglitol	1996	Antidiabetic (Type II)	Bayer
Topotecan	1996	Antineoplastic	SmithKline Beecham
Docetaxel	1995	Antineoplastic	Rhône-Poulenc Rorer
Tacrolimus	1993	Immunosuppressant	Fujisawa
Paclitaxel	1993	Antineoplastic	Bristol-Myers Squibb

Table 1.2 Some of the natural products that are under development for therapeutic use[18]

Natural product	Source	Target	Disease	Status of development
Manoalide	Marine sponge	Phospholipase-A2	Anti-inflammatory	Clinical trials
Dolastatin	Sea hare	Microtubules	Antineoplastic	Non-clinical
Staurosporine	Streptomyces	Protein kinase C	Antineoplastic	Clinical trials
Epothilone	Myxobacterium	Microtubules	Antineoplastic	Research
Calanolide	Tree	DNA polymerase action on reverse transcriptase	Acquired immuno-deficiency syndrome (AIDS)	Clinical trials
Huperzine	Moss	Cholinesterase	Alzheimer's disease	Clinical trials

Examples of such biological activity profiles of natural products would include, nootropics, psychoactive agents, dependence attenuators, anticonvulsants, sedatives, analgesics, anti-inflammatory agents, antipyretics, neurotransmission modulators, autonomic autacoid activity modulators, (vasodilator/constrictor biological factor, which are like local hormones, acting near the site of synthesis), anticoagulants, hypolipidemics, antihypertensive agents, cardioprotectants, positive ionotropes, (increase the strength of muscle contraction) antitussives, (cough suppressants) antiasthmatics, pulmonary function enhancers, antiallergents, hypoglycemic agents, antifertility agents, fertility-enhancing agents, wound-healing agents, dermal healing agents, bone-healing agents, compounds useful in the prevention of urinary calculi as well as their dissolution, gastrointestinal motility modulators, gastric ulcer protectants, immunomodulators, hepato-protective agents, myelo-protective agents, pancreato-protective agents, oculo-protective agents, membrane stabilizers, haemato-protective agents, antioxidants, agents protective against oxidative stress, antineoplastics, antimicrobials, antifungal agents, antiprotozoal agents, antihelminthics, and nutraceuticals.[19]

The importance of natural products derives from their structural diversity, with higher plants producing secondary metabolites of greater molecular diversity than most other classes of organisms. Natural products provide greater structural diversity than any combinatorial or other practicable synthetic approach. Drug discovery has largely overcome the restrictions of working with extracts of natural products by integrating novel solid-phase fractionation and chromatography techniques with screening assays specifically optimized for natural products. These are linked to mass spectrometry and nuclear magnetic resonance techniques for rapid identification of the structures of active compounds.

CLASSIFICATION OF METABOLITES

Metabolites can be classified as primary and secondary metabolites. The former includes proteins, nucleic acids and carbohydrates which are essential and utilized for the existence of the organism (plant/marine). The secondary metabolites like flavonoids, terpenoids, etc. that are needed for their survival are more utilized in their inter-organism interactions.

The relation between the primary and secondary are illustrated in Figure 1.4. In the biosynthesis of secondary metabolites there exist four principle building blocks, viz., acetate, mevalonate, shikimate and amino acids. However, most of these intermediates are also used for the production of primary metabolites such as the fatty acids, nucleosides and polypeptides.

A theory therefore has been put forward that suggests that the secondary metabolism arose as a way of using the acetate, shikimate and amino acids that were excess to requirements ("shunt metabolites"). It was proposed that changes in the genetic make-up of an organism (mutations), whether they be natural chemical mistakes or exposure to chemicals, viruses or radiation, have lead to the increased production of the key intermediates. When these compounds are able to provide useful biological activities that increase the chances of survival for an organism, they are more likely to be passed on through the generations, hence, Darwin's famous theory on evolution through "survival of the fittest" comes into play. This is particularly evident when comparing higher organisms such as humans and lower organisms that live in extreme ecological niches. Humans produce few secondary metabolites since our survival is determined by physical means. In contrast, the lower organisms such as fungi and bacteria, which generally live in highly competitive environments, are considered as major producers of secondary metabolites.

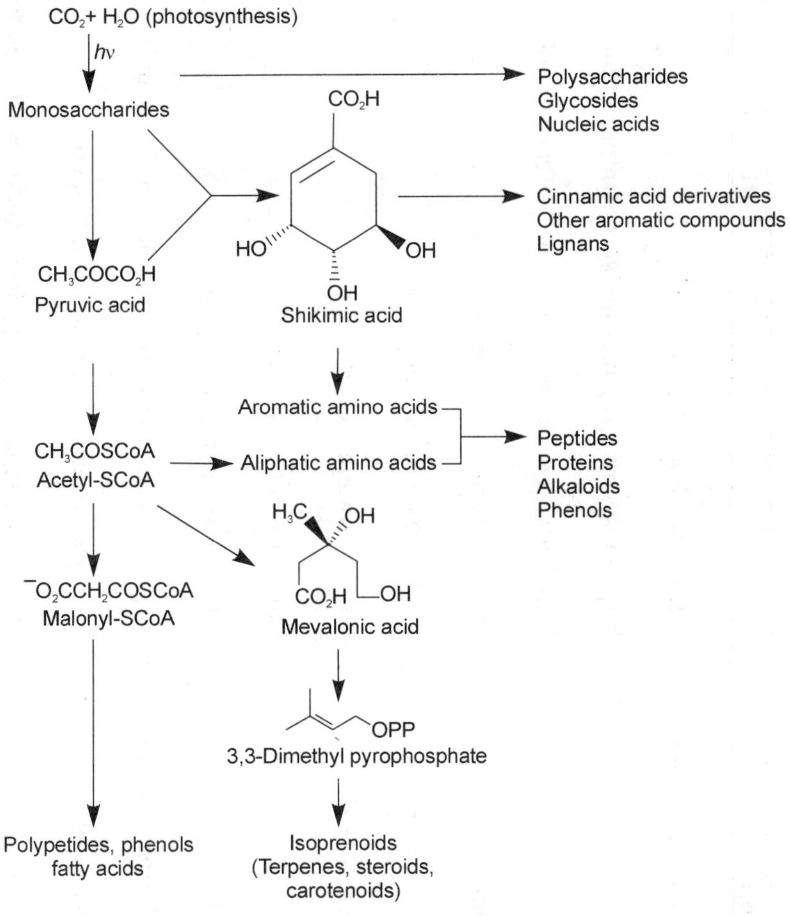

Figure 1.4 Primary metabolites and their links to secondary metabolism

INFORMATION SOURCE OF NATURAL PRODUCTS

Literature survey forms the primary and the most important protocol in any research. Certain plants are unique to geography, climate, soil, etc. Research in the area of natural products can be followed in the journals enlisted in Table 1.3, and given along with their internet address.

Table 1.4 gives the internet resources for natural products information.[20]

Table 1.3 Journals on natural products[20]

Description	URL
American Journal of Natural Medicine	http: //www.impakt.com
Botanical Review	http: // www.nybg.org/bsci/spub/botr/frntpg3b.html
Bulletin on Narcotics	http: // www.odccp.org/bulletin_on_narcotics.html
Canadian Journal of Botany	http: // www.herbalist.on.ca/journal/
Economic Botany	http: // www.econbot.org
European Journal of Herbal Medicine	http: // www.ejhm.co.uk/
Herb Companion Press	http: // www.interweave.com
Herb Quarterly	http: // www.herbquarterly.com
International Journal of Aromatherapy	http: // www.harcourt-international.com/journals/ijar
Journal of Ethnopharmacology	http: // www.elsevier.com/locate/jethpharm
Journal of Natural Products	http: // www.pubs.acs.org/journals/jnprdf/index.html
Medical Anthropology: Cross Cultural Studies in Health and Illness	http: //www.sfu.ca/medanth

Description	URL
Medical Herbalism: A Journal for the Clinical Practitioner	http: // medherb.com/MHHOME. SHTML
Natural Health	http: // www.naturalhealth 1.com
Natural Product Letters	http: // www.tandf.co.uk/journal/titles/10575634.html
Natural Product Reports	http: // www.rsc.org/is/journals/current/npr/nprpub.htm
Pharmaceutical Biology	http: // www.szp.swets.nl/szp/frameset.htm
Phytochemistry: The International Journal of Plant Biochemistry and Molecular Biology: The Journal of the Phytochemical Society of North America	http: // www. elsevier.nl/locate/inca/273
Phytotherapy Research	http: // www3.interscience.wiley.com/cgi-bin/jtoc?ID=12567
Plant Foods for Human Nutrition	http: // www.wkap.nl/jrnltoc.htm/0921-9668/contents
Planta Medica: Natural Products and Medicinal Plant Research	http: // www.thieme.de/plantamedica/fr_inhalt.html
Toxicon: Official Journal of the International Society on Toxicology	http: // www. elsevier.com/locate/toxican
Z Naturforsch	http: // www.znaturforsch.com/

Table 1.4 Internet address for websites carrying natural products information

Description	URL
American Botanical Council	http: //www.herbalgram.org
American Herbalists Guild	http: // www.americanherbalistsguild.com/top.htm
Complementary & Alternative Methods	http: // www.cancer.org/eprise/main/docroot/eto_5?sitearea = eto
Dr. Duke's Phytochemical and Ethnobotonical Databases	http: // www.ars-grin.gov/duke
Herb Research Foundation	http: // www.herbs.org
Herbal Education Services	http: // www.botanicalmedicine.org
Herbal Medicine: Internet Resources: Alternative Medicine	http: // www.pitt.edu/~cbw/herb.html
HerbMed	http: // www.herbmed.org
International Herb Association	http: // www.iherbs.org
MEDLINEplus: Alternative Medicine	http: // www.nlm.nih.gov/medlineplus/alternativemedicine.html
National Center for Complementary and Alternative Medicine	http: // www.nccam.nih.gov/

Description	URL
World Health Organization Publications	http: // www.who.int/dsa/cat98/trad8.htm
AGRICOLA (AgriculturalOnline Access)	http: // www.nal.usda.gov/general_info/agricola/agricola.html
American Herbal Pharmacopoeia	http: // www.herbal-alp.org
Ayurvedic Medicines	http: // www.dabur.com www.thehimalayadrugco.com
Chinese Medicine	http: // www.cintcm.com/index.htm
Facts and Comparisons, The Review of Natural Products	http: // www.factsandcomparisons.com
Herbs, Chemistry	http: // fridele.com http: // realtime.net/anr http: // www.aspp.com
Natural Medicines Comprehensive Database	http: // www.naturaldatabase.com
World Health Organization (WHO)	http: // who. Int/medicines/library/trm/medicinalplants/monographs.shtml

IMPACT OF NATURAL PRODUCTS ON HUMAN SOCIETY

Natural products are intertwined with the lives of human beings be it in the form of flavouring agent or colouring material or fragrances or medicine or pesticides. Life would lose its flavour, but for natural products.

The presence of natural products resulting in properties ranging from antioxidants to pesticides are listed in Table 1.5.

Table 1.5 List of natural products and their properties

Potential uses	Compounds	Typical sources
Antioxidants	Carnosic acid	Rosemary, sage
	Flavonoids	Grape seeds, tea
	β-Carotene	Carrot, papaya
Colouring agents	Anthocyanins	Cornflowers, grapes
	Xanthophylls	Orange peel, corn
	Chlorophyll	Wheat grass
Flavouring agents	Citrus oil	Orange, lemon
	Ginger oil	Ginger
	Vanillin	Vanilla beans
Emulsifiers and stabilizers	Lecithin	Soybean, canola
	Aloe gel	Aloes
Pharmaceuticals	Taxol	Yew tree
	Ginsenosides, ginkgolides,flavonoids	Panax ginseng
	Genistein, daidzein	Ginkgo Kudzu, Soy
	Hyperforin, hypericin	St.John's wort
Fragrances	Jasmine oil	Jasmine
	Rose oil	Rose
	Lavender oil	Lavender
Specialty oils and fatty acids	Jojoba oil	Jojoba
	Sunflower oil	Sunflower
	Gamma linolenic acid (GLA)	Evening primrose
Pesticides	Azadirachtin	Neem seeds
	Pyrethrins	Chrysanthemum
	Nicotine	Tobacco

Higher plants produce a variety of different types of compounds, including biologically active proteins. Many polypeptides, which exhibit interesting properties[21] have been isolated from the venoms of arachnids and arthropods that prey on insects. Scorpion venom has been shown to affect ion channel and also cardiovascular system. The caterpillar *Lonomia achelous* has been reported to be the source of a biologically active protein that causes a coagulopathy that is mediated via specific interactions with Factor V in the coagulation cascade which may pave way for thrombosis research.

Marine natural products (Figure 1.5) have engaged the attention of researchers for the past five decades. Till date, 17,000 natural products have been isolated and reported in over 7000 publications.

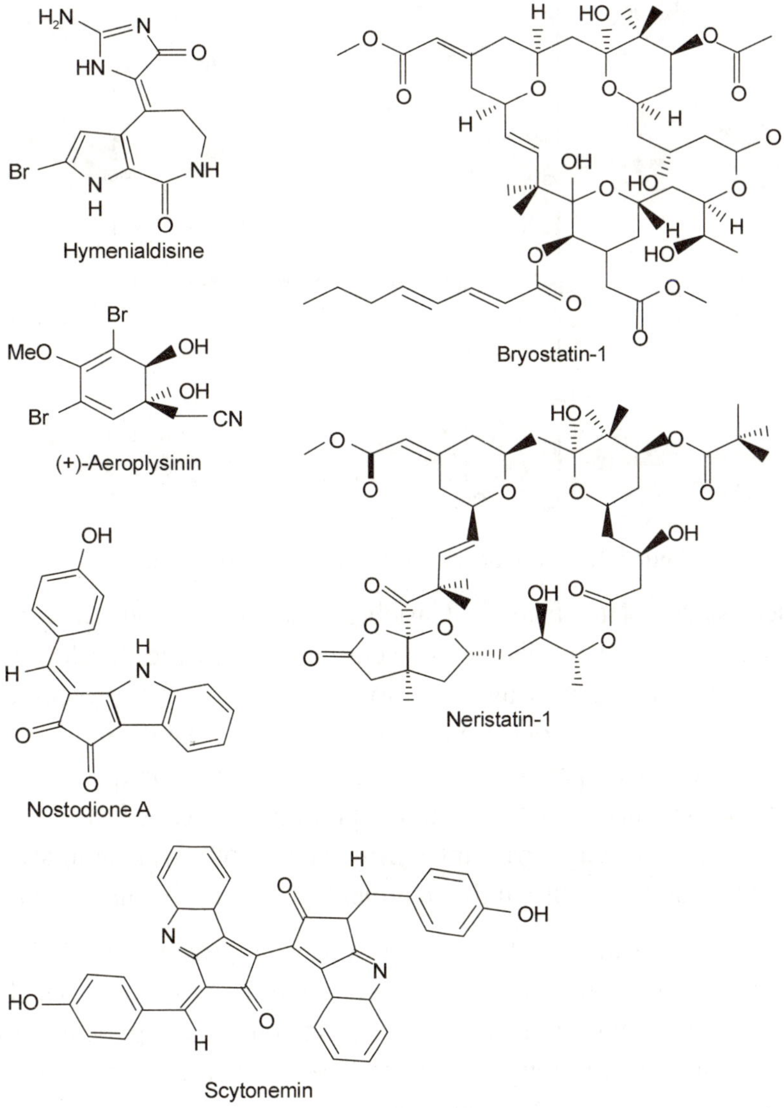

Figure 1.5 Some phytochemicals of marine origin

The source of these compounds includes algae, fungi, bacteria, invertebrates, etc. Natural products from marine organisms are released into water and therefore are rapidly diluted, (accordingly they must be very potent materials to have the desired end effect.)

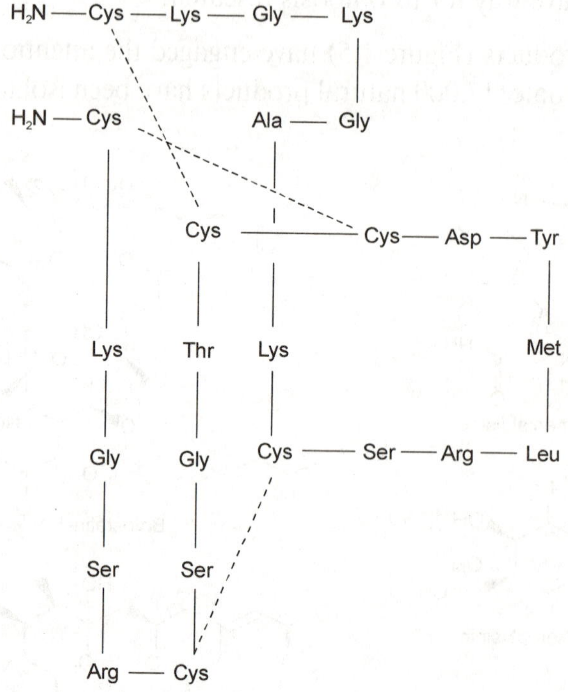

Figure 1.6 Conotoxin from *Conus magus* (cone snail)

The C-nucleosides isolated from the Caribbean sponge *Cryptotheca crypta* were found to possess antiviral activity. This discovery eventually led to the development of cytosine arabinoside, a useful antineoplastic agent. Biologically active marine proteins derived from the venom of marine snails of the *Conus* genus have attracted a significant level of research over the years. [21] These conotoxin peptides (Figure 1.6) interact in a unique fashion with voltage-gated ion channels to induce a wide spectrum of pharmacological effects. Such effects include anaesthesia, analgesia, and anticonvulsant activity. The conotoxin ziconotide is currently under review in the United States for use in the treatment of chronic, opiate-resistant pain.

According to some estimates, there are most likely approximately 1000 different *Conus* snails. Each snail produces up to approximately 200 different venoms. The broad spectrum of biological activities manifested by each of these venom components multiplied by the number of snails and venom components available, suggests significant opportunity for new drugs from the snail alone. Marine sponges have also been reported to be a source of halogenated compounds (Figure 1.7).

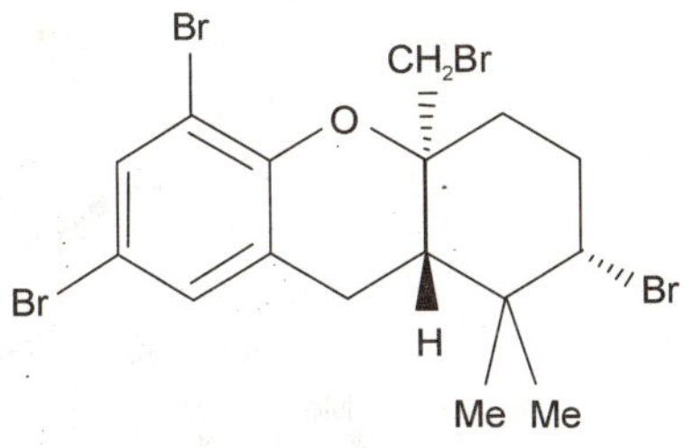

Figure 1.7 A xanthone derivative isolated from marine sponge *Scalarispongia scalaris*

Marine Medicines[22]

Ecteinascidia turbinata, a Caribbean sea squirt, makes a compound (branded as Yondelis). Awaiting approval as antitumour compound for soft-tissue sarcomas and also for ovarian cancer.

Conus magus, a cone snail, paralyses its prey using a poison tipped barb. The poison is a painkiller many times more potent than morphine, and is now on the market as Prialt.

WHAT MAKES THE NATURAL PRODUCTS "WHAT THEY ARE?"

Every chemical molecule that exerts biological or pharmacological activity possess some essential feature(s) in their chemical structure that makes their activity specific. For example, all molecules that target the brain (be it psychotic or a stimulant like caffeine) possess a nitrogen atom—mostly in their ring, to penetrate the blood–brain barrier. A few of such features are given below.

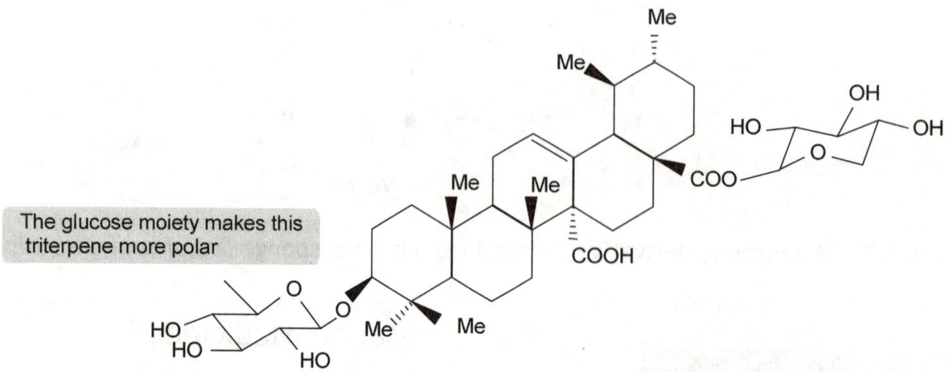
The glucose moiety makes this triterpene more polar

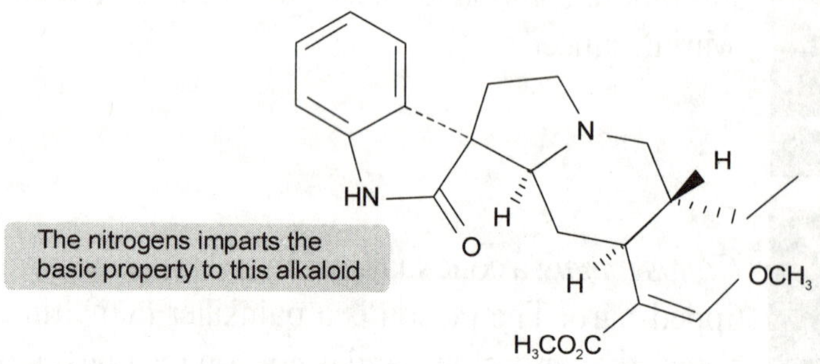
The nitrogens imparts the basic property to this alkaloid

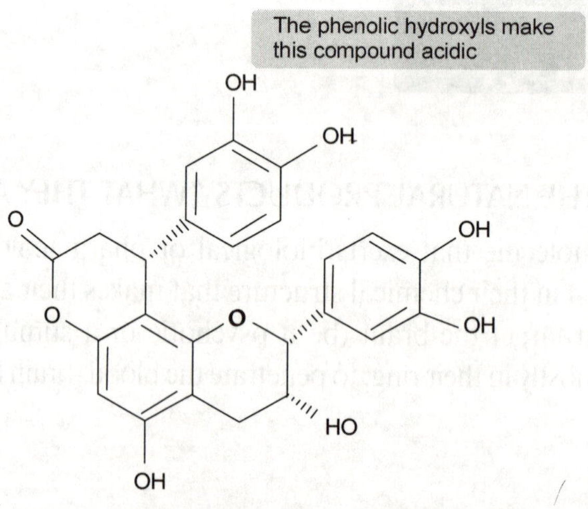
The phenolic hydroxyls make this compound acidic

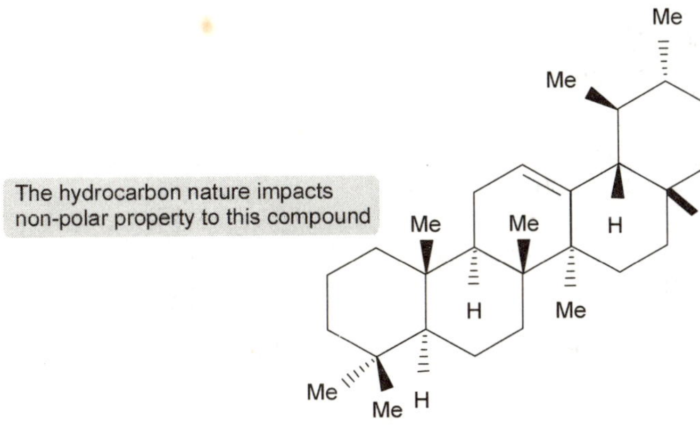

STRUCTURAL DIVERSITY—AN ILLUSTRATION

The importance and complexity in the world of natural products lies in the diversified structural variation of the chemicals obtained from them. Molecules possessing around sixty atoms are termed as small molecules and more than that are known as macromolecules. Proteins and DNA belong to this macromolecule category. The structures given below is an illustration of such variation for molecules obtained from different sources.

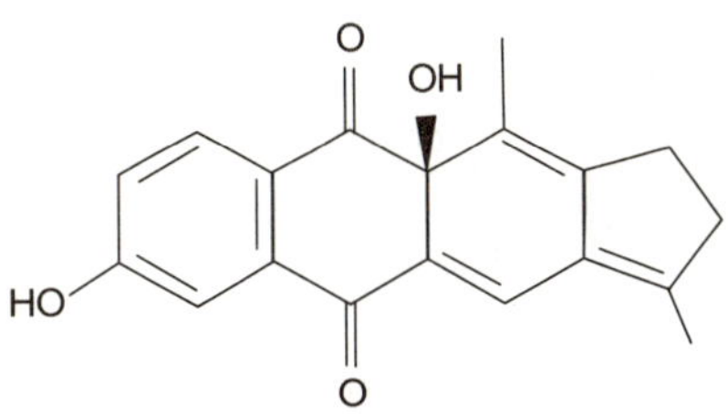

Isokigelinol from Bignoniaceae from *Salvia sclarea*

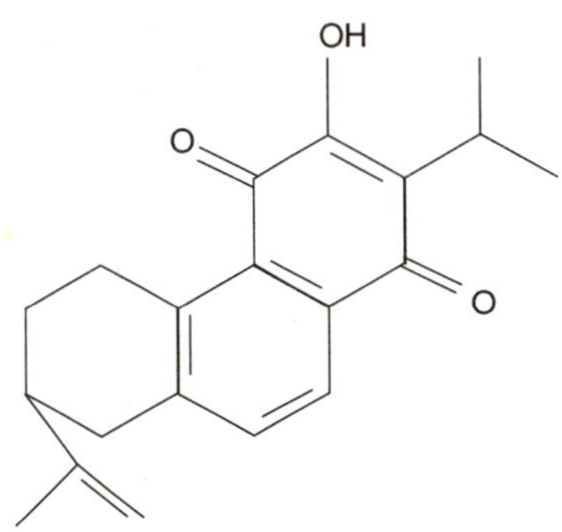

Dehydrosalvipisone

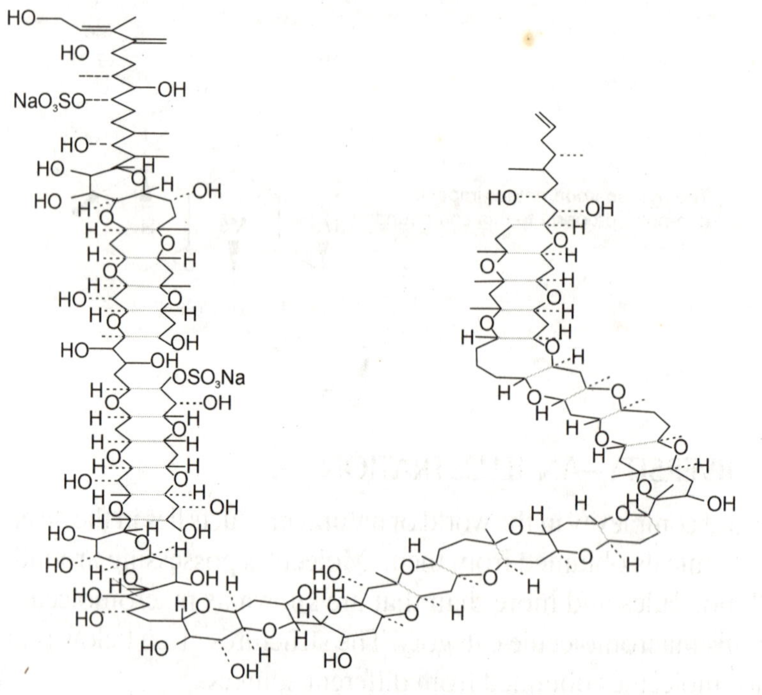

Manzamine A—an alkaloid from marine sponge

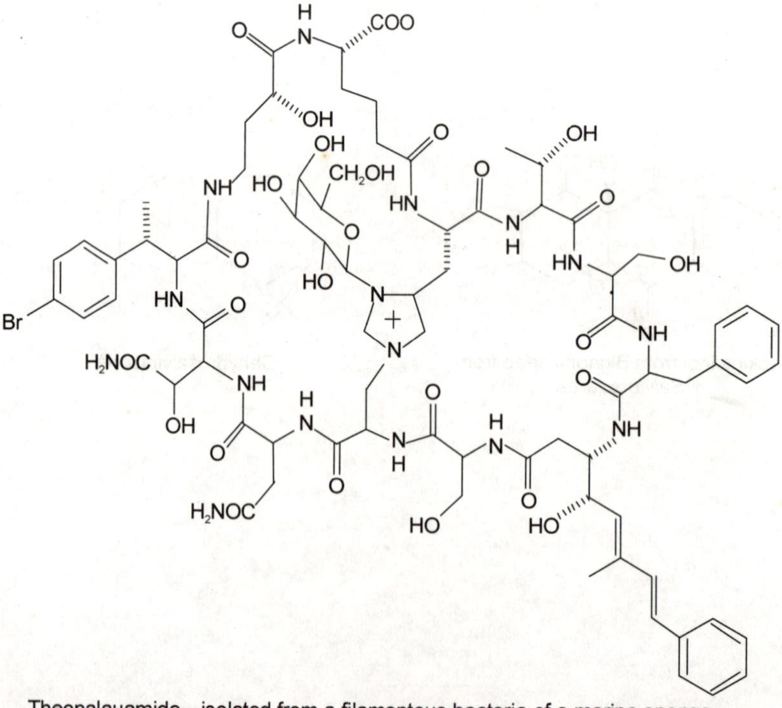

Theopalauamide—isolated from a filamentous bacteria of a marine sponge

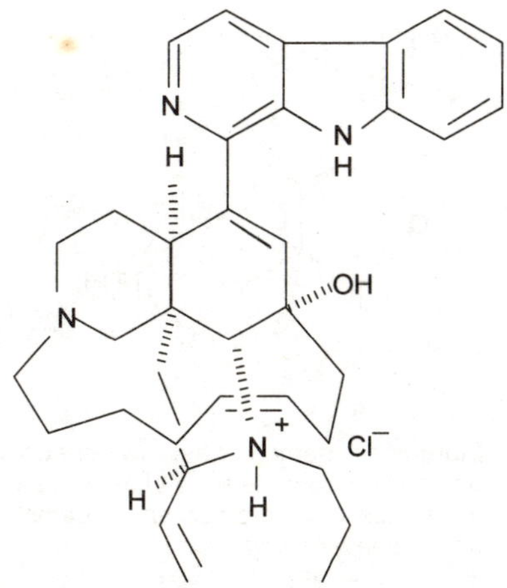

Maitotoxin isolated from Gambierdiscus a toxin

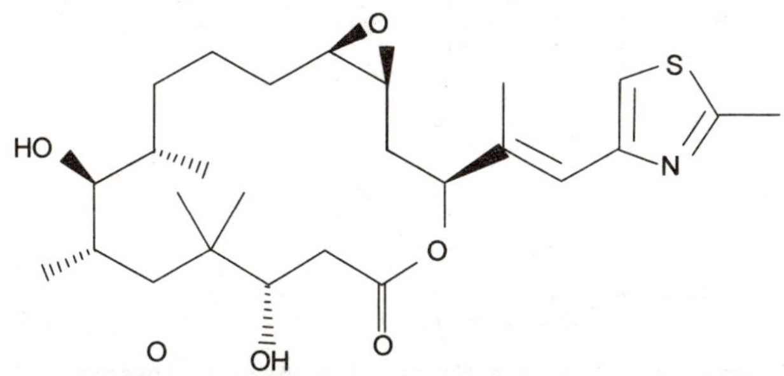

Epothilones A from a culture of the toxicus myxobacterium, *Sorangium cellulosum*

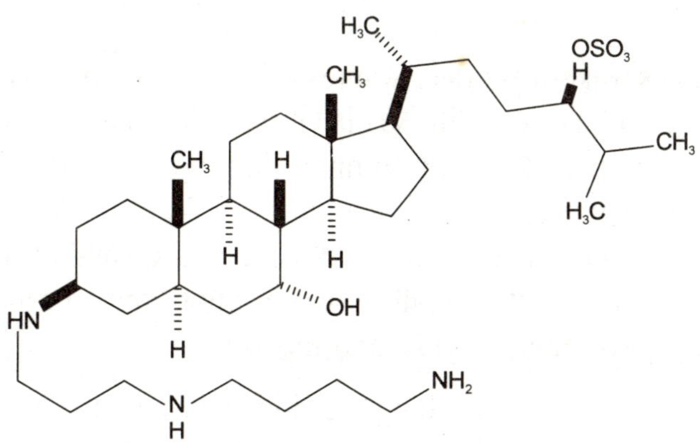

Squalamine from *Squalus acanthias*

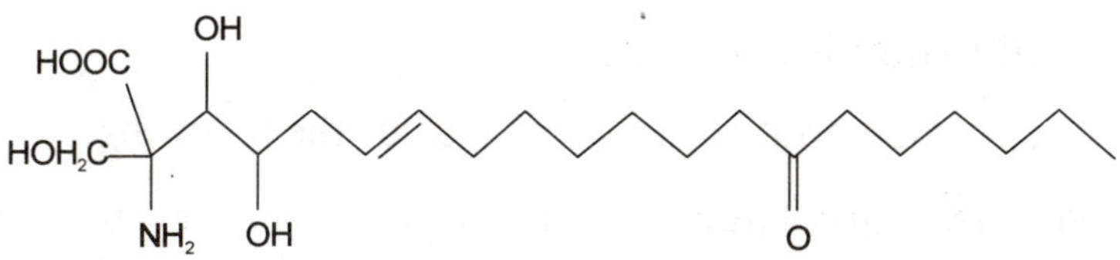

Myriocin isolated from the fungus (dogfish shark) *Myriococcum albomyces*

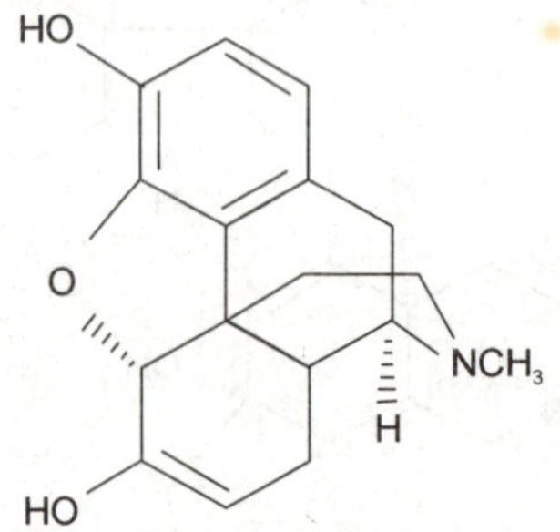

Morphine (named after Morpheus, antifeedant from the seed of neem tree the Greek God of sleep and dreams) from *Papaver somniferum*

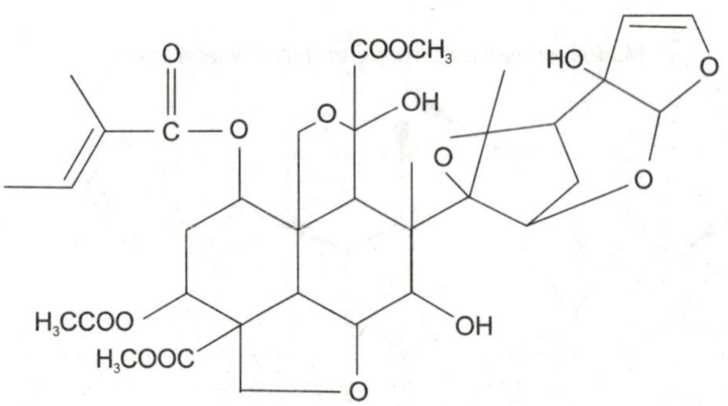

Azadirachtin, an insect antifeedant, from the seed of neem tree

EXTRACTION

Extraction is what brings us wake-up coffee. Solutes within the powdered raw material move or partition into the solvent phase and diffuse out of the solid matrix and eventually out of the particulate bulk. It is important to keep in mind that the isolation of natural products differs from that of the more prevalent biological macromolecules. This is because natural products are typically secondary metabolites and as such are smaller in size, chemically more diverse in structure, and present in smaller concentrations than the more homogeneous proteins, carbohydrates, lipids, nucleic acids, and the like.

While planning for an extraction, two things need to be taken into consideration, namely the completeness of the extraction and obtaining the intended compound without degradation.

Traditional Solvent Extraction Techniques

The techniques for extraction normally followed include the following.

Steam distillation Steam distillation is the method of choice for extraction of essential oils which form the base of the perfumery industry. Direct or indirect steam can be used.

The steam is passed onto the flowers and it carries away the volatile components which are then cooled. Recovery of the essential oil is facilitated by distillation of two immiscible liquids, namely, water and essential oil, based on the basic principle that, at the boiling temperature, the combined vapour pressure equals the ambient pressure. Thus the essential oil ingredients whose boiling points normally range from 200 to 300°C, are made to boil at a temperature close to that of water. They remain as oily substance on the top of the aqueous layer and can be separated easily.

Maceration *Maceration*[23] is one of the primitive methods used for the recovery of "absolute" or essential oil from flowers. "Absolute" describes an aromatic oil in the liquid or semi-liquid states. The oil sacs of the fragrant flowers or botanical matter are ruptured by immersing them in a molten fat or oil at 60–70°C, which in turn absorbs the aroma of the flowers or the substrate. Spent flowers are separated from the hot fat or oil, and the residual fat or oil is removed by hydraulic press. The treated fat is reused for the next batch of fresh flowers or botanicals until it is saturated with aroma. This perfumed pomade is either marketed as such or subsequently extracted with alcohol to produce "absolute." But maceration is a very tedious, time-consuming, and inefficient method, used in the past for jasmine flowers.

Normal percolation This involves soaking the bark/seeds overnight in appropriate solvent at room temperature and the solvent then distilled off to obtain an extract. This is repeated twice or thrice for complete extraction. A typical extraction protocol is given below in Figure 1.8.

Refluxing the material Since temperature and solubility of compounds are directly proportional, by refluxing the material with the solvent at a particular temperature, complete extraction is possible. But this is detrimental for compounds that are sensitive to heat as this degrades them.

Soxhlet extraction effects the usage of less amount of solvent to extract more quantity of the metabolite mixture. In this process, the vapours of the solvent are allowed to interact with the material to be extracted and then condensed and the solvent alone is vaporized to repeat the extraction. This results in a complete extraction.

Microwave-mediated extraction of essential oils has helped in maintaining the mild conditions and effects superior extraction.

Ultrasonic extraction Although the use of ultrasonic energy to aid the extraction of medicinal compounds from plant materials can be found in literature as early as the 1950s, mechanistic aspects of the usefulness of ultrasonically assisted extraction are worth noting. Fundamentally, the effects of ultrasound on the cell walls of plants can be described as follows:

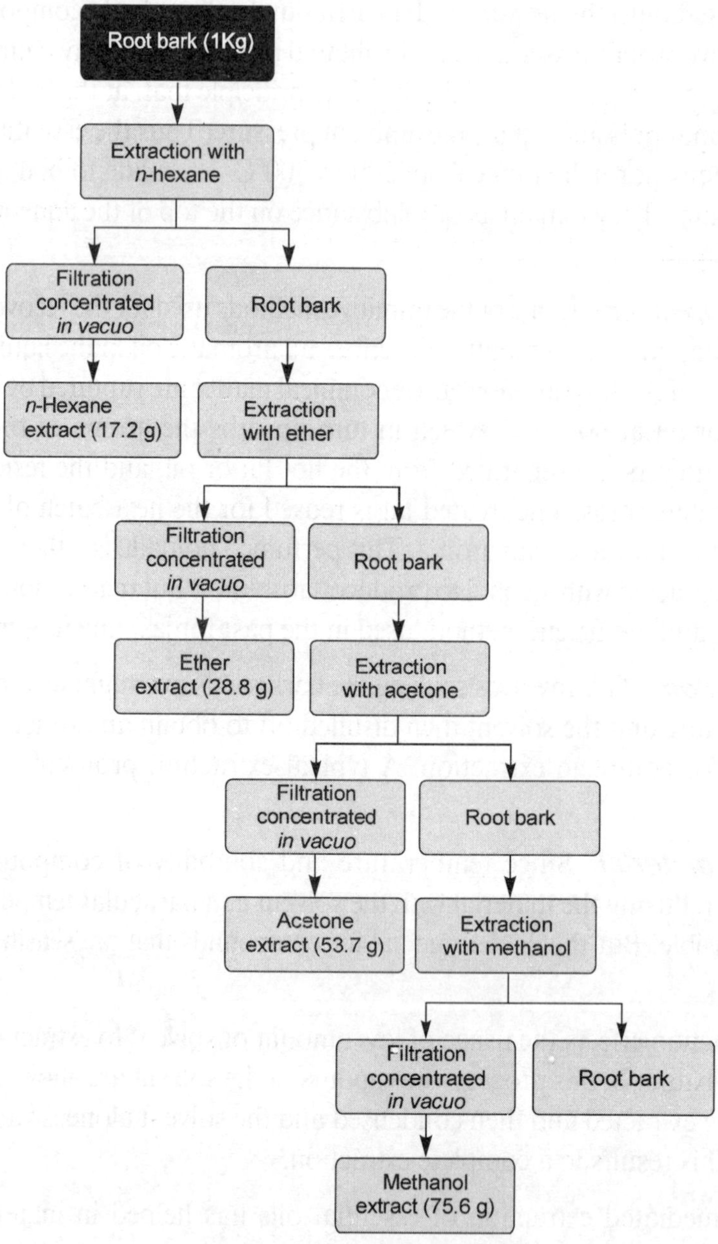

Figure 1.8 An extraction protocol

1. Some plant cells occur in the form of glands (external or internal) filled with essential oil. A characteristic of external glands is that their skin is very thin and can be easily destroyed by sonication, thus facilitating release of essential oil contents into the extraction solvent.

2. Ultrasound can also facilitate the swelling and hydration of plant materials to cause enlargement of the pores of the cell wall. Better swelling will improve the rate of mass transfer and, occasionally, break the cell walls, thus resulting in increased extraction efficiency and/or reduced extraction time.

Solvent extraction Solvent extraction is usually used to recover a component from either a solid or liquid. The sample is contacted with a solvent that will dissolve the solutes of interest. Solvent extraction is of major commercial importance to the chemical and biochemical industries, as it is often the most efficient method of separation of valuable products from complex feedstock or reaction products. Some extraction techniques involve partition between two immiscible liquids, others involve either continuous extractions or batch extractions. Because of environmental concerns, many common liquid/liquid processes have been modified to either utilize benign solvents, or move to more frugal processes such as solid phase extraction. The solvent can be a vapour, supercritical fluid, or liquid, and the sample can be a gas, liquid or solid.

Solid phase extraction Solid phase extraction (SPE) is an alternative to liquid/liquid extraction, and has become the method of choice for the separation and purification of a wide range of samples in the laboratory. The sample is usually dissolved in an appropriate solvent and passed through a small bed of adsorbent of very consistent particle size and shape to maximize separation efficiency. The compounds are eluted with step changes of small volumes of solvents. The major advantage is that solvent volumes are greatly reduced. There is a newer, modified technique that is used in analytical laboratories, called solid phase micro extraction. This immerses a fused silica fibre coated with a stationary phase into the sample solution for several minutes.

Enfluerage method of extraction from flowers The traditional cold-fat extraction process is known as " enfleurage". It is a very interesting, historical process used to obtain the essential oils and perfume components from rose, jasmine, and other flowers. The rose and jasmine flowers continue to produce perfume during the long process. Thus the technique can obtain more perfume from those flowers than if they were just macerated and extracted by hot fat, solvent or steam when they were picked — as happens to many other plant perfume sources. The process uses a fat comprised of 40 parts of beef tallow and 60 parts of lard. The two fats are melted together, and repeatedly beaten under cold water and alum solutions to purify them. Benzoin is added to the fat mixture to prevent biological degradation.

The fat is spread to about 4 mm thickness on both sides of 0.5 × 0.5m glass plates in wooden frames. Flowers are pressed into the fat on one side of the frame only, and the frames stacked vertically so that the flowers are very close to the layer of fat on the frame above. After

1–3 days, the flowers are stripped off and fresh flowers added to the other layer of fat that had not been used, and the frames are again stacked. The cycle is repeated about 30–35 times, or until the fat is saturated with perfume. The saturated fat is known as "pomade". The fat is removed from the frames and extracted with alcohol to collect the perfume.The alcohol is cooled and filtered to remove most of the dissolved fat. The alcohol solution is called the "extract", and the residue after evaporation of the solvent is known as the "enfleurage absolute".

Supercritical Fluid Extraction

Over the last two decades as the alternative to the traditional solvent extraction of natural products, supercritical fluid extraction technique has emerged as a superior technique. Supercritical fluid extraction is the process of isolating a component from a matrix, using supercritical fluids. It uses a clean, safe, inexpensive, nonflammable, nontoxic, environment-friendly, nonpolluting solvent, such as carbon dioxide. Besides, the energy costs associated with this novel extraction technique are lower than the costs for traditional solvent extraction methods.

Supercritical fluid extraction technology is thus increasingly gaining importance over the conventional techniques for extraction of natural products.

Supercritical fluid (SCF) is a state of a substance where it retains the properties of both a liquid and a gas. This is the state wherein the maximum solvent capacity and the largest variations in solvent properties can be achieved with small changes in temperature and pressure. It offers very attractive extraction characteristics, owing to its favourable diffusivity, viscosity, surface tension and other physical properties. Its diffusivity is one to two orders of magnitude higher than those of other liquids, which facilitates rapid mass transfer and faster completion of extraction than conventional liquid solvents.

Its low viscosity and surface tension enable it to easily penetrate the botanical material from which the active component is to be extracted. The gas-like characteristics of SCF provide ideal conditions for extraction of solutes giving a high degree of recovery in a short period of time. However, it also has the superior dissolving properties of a liquid solvent. It can also selectively extract target compounds from a complex mixture. Sometimes the target compound is the active ingredient of interest. At other times, it may be an undesirable component which needs to be removed from the final product. The strong pressure and temperature (or density) dependence of solubility of certain solutes in an SCF solvent is the most crucial phenomenon that is exploited in supercritical fluid extraction (SCFE). Many of the same qualities which make SCF ideal for extraction, also make them good candidates for use as a superior medium for chemical reactions offering enhanced reaction rates and preferred selectivity of conversion. Once such a reaction is over, the fluid solvent is vented to precipitate the reaction product. A comparison of physical properties of some SCF solvents is given in Table 1.6.

Table 1.6 Physical properties of some common solvents used in SCF state[24]

Fluid	Normal boiling point (°C)	Critical constants		
		Pressure (bar)	Temperature (°C)	Density (g/cm^3)
Carbon dioxide	−78.5	73.8	31.1	0.468
Ethane	−88.0	48.8	32.2	0.203
Ethylene	−103.7	50.4	9.3	0.20
Propane	−44.7	42.5	96.7	0.220
Propylene	−47.7	46.2	91.9	0.23
Benzene	80.1	48.9	289.0	0.302
Toluene	110.0	41.1	318.6	0.29
Chlorotrifluoromethane	−81.4	39.2	28.9	0.58
Trichlorofluoromethane	23.7	44.1	196.6	0.554
Nitrous oxide	−89.0	71.0	36.5	0.457
Ammonia	100.0	112.8	132.5	0.240
Water	100.0	220.5	374.2	0.272

The most desirable SCF solvent for extraction of natural products for food and medicines today is carbon dioxide (CO_2). It is an inert, inexpensive, easily available, odourless, tasteless, environment-friendly, and GRAS (generally regarded as safe) solvent. Further, in SCF processing with CO_2, there is no solvent residue in the extract, since it is a gas in the ambient condition. Also, its near-ambient critical temperature (31.1°C) makes it ideally suitable for thermolabile natural products. Due to its low latent heat of vaporization, low energy input is required for the extract separation system which renders the most natural smelling and natural-tasting extracts.The solubility of some naturally occuring organic molecules in supercritical carbon dioxide are given in Table 1.7.

Table 1.7 Botanical ingredients and their solubility in liquid and supercritical carbon dioxide

Very soluble	Sparingly soluble	Almost insoluble
Nonpolar and slightly polar low M.W. (<250) organics, e.g., Mono-and sesquiterpenes, thiols, pyrazines, and thiazoles, acetic acid, benzaldehyde hexanol, glycerol, acetates	Higher M.W. organics, (<400), e.g., substituted terpenes and sesquiterpenes, water, oleic acid, glycerol, decanol, saturated lipids up to C_{12}	Organics with M.W. above 400, e.g., sugars, proteins, tannins, waxes, inorganic salts, chlorophyll, carotenoids, citric, malic acids, amino acids, nitrates, pesticides, insecticides, glycine, etc.

SCFE applications in natural products and food industries The commonly applied arena of SCF includes decaffeination of coffee and tea, spice extraction (oil and oleoresin), deodorization of oils and fats, extraction of vegetable oils from flaked seeds and grains, hops extraction for bitterness, extraction of herbal medicines, extraction of lanolin from wool, deoiling of fast foods, decholesterolization of egg yolk and animal tissues, extraction of antioxidants from plant materials, food colours from botanicals, and natural pesticides, and denicotinization of tobacco.

As can be seen from the above applications, it is not only the effectiveness but also the selectivity that marks this technique greener.

The drive behind isolation of natural products varies from scientist to scientist. A chemist would be interested in looking into the chemical structure of the compound, a biologist into the bioactivity, a medicinal chemist into its synthetic feasibility and so on. Before planning for an extraction, one should know as to what type of compound is to be isolated. Is it a polar, nonpolar, acidic or basic compound. But the commonly followed approach is bioactivity-guided isolation. The material of interest (bark/root/leaves) is extracted in different solvents of varying polarity, say, hexane extract, chloroform extract and methanolic extract. Then each extract is tested for its bioactivity—antifungal, antifeedant, etc. The one that shows promising result is then taken for further studies.

Bioassay and the Chemistry of Natural Products

Bioassay is a technique for screening a compound(s) for a particular biological or pharmacological activity. They are the foundation of a natural products discovery group. The bioactivity can include antifungal, antiviral, anti-feedant, anticancer, etc. It is very difficult for any research group to possess the required manpower, financial resources, or time to extract every compound from an extract. Bioassays permit the researchers to prioritize their investigations. Bioassays can be used in three ways.

The first way is used to determine if the extracts are active (prescreening). As the chance that an extract is active is quite low, large numbers of extracts should be tested. This is usually performed at a specific concentration (dose) previously determined to be "interesting". If the extract responds positively, it will be subjected to further testing.

The second manner (screening) permits the prioritization of the active extracts. The extracts are tested over a range of concentrations and their responses quantified (i.e., an IC_{50} is assigned which is the concentration required to inhibit the growth of a microorganism by 50%). More active fractions should be investigated first with greater resources than a less active extract.

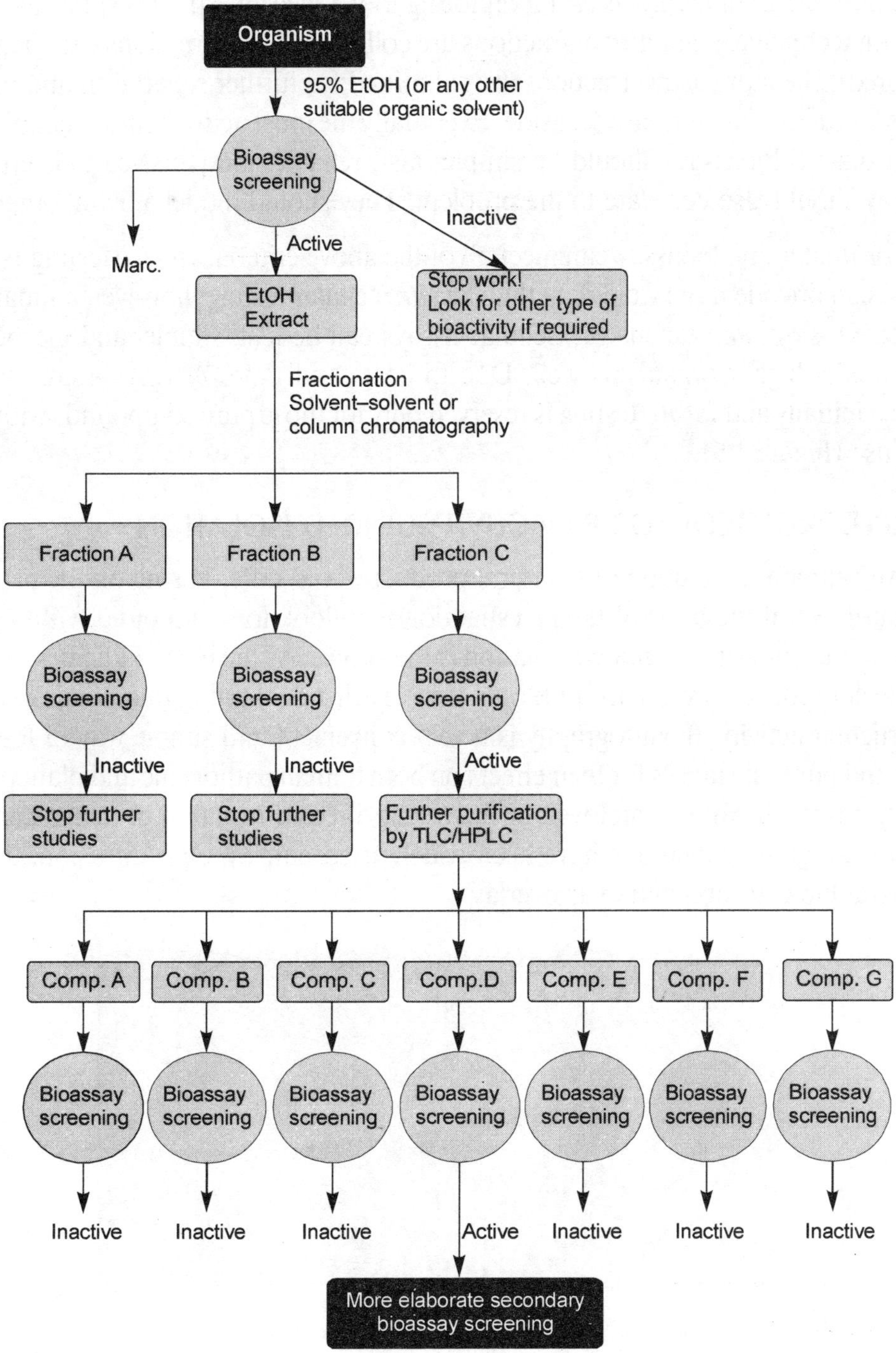

Figure 1.9 The bioassay protocol

The third use of bioassays is as a monitoring tool. Once an extract is subjected to a given separation technique, a number of fractions are collected. These fractions can then be tested (monitored); the more active fractions are submitted for further separation and monitoring whereas less active fractions are set aside. Eventually the most active component of an extract will be isolated. Bioassays should be simple, fast, reliable, inexpensive, and reproducible. The assay should also correlate to the problem. They should model a living organism well.

Unfortunately, no bioassay can meet all of the above criteria. *In vivo* testing (such as on animals) can provide more valid data than *in vitro* cellular testing; however, animal testing is complicated, slow, and expensive. Cellular assays can be fast, simple, and inexpensive but do not model higher organisms well. Due to costs and time, *in vitro* assays are usually employed initially and *in vivo* testing is reserved only for those pure compounds with potential clinical use (Figure 1.9).

A SAMPLE PROTOCOL FOR BIOACTIVITY-GUIDED ISOLATION

The drive behind the isolation of a natural product would differ basing on the nature of the investigator. But if the basis of the investigation is to look for compound with a particular bioactivity, then bioactivity-guided isolation is the choice. As the name indicates, the attempt to isolate compounds are carried out, only if the extract containing that compound exhibits that particular activity. Bioautography is a very convenient and simple way of testing plant extracts and pure substances for their effects on both human pathogenic and plant pathogenic microorganisms. It can be employed in the target-directed isolation of active constituents. Three bioautographic methods have been widely used namely, agar diffusion, direct TLC bioautographic detection and agar overlay.

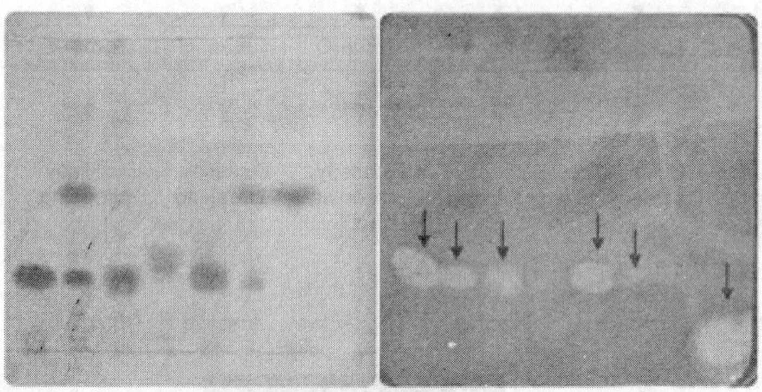

Plate A Plate B

Plate A, given above is the elution and visualization of extracts in a pre-coated thin layer chromatographic plate. Each spot, corresponds to a single compound. Plate B is the result after elution and spraying with a fungus. The white spots (indicated by the arrow) are the compounds that show resistance to the fungal growth. Now if the aim of the work is to isolate some antifungal compounds, then only those showing the activity here needs to be isolated.

A typical isolation protocol is given below.

Dried and milled fruits of *Coutarea hexandra* were extracted twice in 1L of 80% ethanol at room temperature by stirring overnight. The plant material was filtered off and the ethanol extract combined and evaporated *in vacuo* to leave a marrow residue (18 g, 3.79%). The ethanol extract was subjected to solvent partitioning. Then four partitions of the extract were obtained (hexane, 90% methanol, ethyl acetate and water). The cytotoxicity activity was retained in the 90% MeOH and EtOAc fractions, respectively. The 90% MeOH fraction (0.98 g) was chromatographed on silica gel (5–60 cm) using a gradient of toluene–EtOAc (0 to 100%), followed by an EtOAc–MeOH (60 : 3), EtOAc–MeOH (50 : 50) and MeOH 100%, for the obtained 30 fractions of 125 mL, which according to their TLC profiles were pooled into 23 fractions denominated MF-1 to MF-23. The fraction MF-16 on precipitation with MeOH yielded 23,24-dihydrocucurbitacin F (3) (29 mg, 0.161%), while the fraction MF-20 on precipitation with $CHCl_3$–MeOH (1 : 1), gave 23,24-dihydro–25-acetylcucurbitacin F (4) (5 mg, 0.027%).[25]

QUERIES THAT NEED TO BE ADDRESSED WHILE PLANNING FOR AN EXTRACTION

1. Isolation Target

Whether the isolation target is

1. an unknown compound associated with a particular biological activity,
2. a previously known compound present in a specific organism,
3. a group of compounds within an organism that are all structurally related to each other,
4. all of the metabolites produced by one natural product source that are not produced by another closely related source,
5. all of the molecules of a particular organism.

2. Aim of Isolation

Possible reasons for carrying out an extraction might be

1. the generation and supply of larger amounts of an already known compound so that more extensive biological testing such as pharmacology and toxicology can be performed on the material.

2. the purification of a small amount of material for initial biological and chemical characterization to be performed.
3. to purify sufficient material in order to conduct complete structural studies and further biological activity characterization.

3. Degree of Purity of the Compound to be Isolated

1. If a natural product compound is to be used for biological testing, it is important to know not only the degree of purity of the material but also the nature of the impurities. It needs to be appreciated that the impurities themselves can contribute significantly to any biological activity observed in the screening program.
2. If the material is to be used in more refined pharmacological or pharmacokinetic testing, then the material should generally be at least 99 per cent pure. If, on the other hand, the material is to be used only for chemical characterization, the acceptable level of purity can range from 95 to 99 per cent. Such a range of purity will generally be sufficient for the determination and assignment of a complex chemical structure via such techniques as NMR spectroscopy, infrared (IR) spectroscopy, and MS/MS spectrometry.
3. X-ray diffraction study of the sample would require 99.9 per cent purity, since such purity is the criterion for a compound to crystallize.
4. Investigation using ultraviolet spectroscopy would tolerate purities down to a level of 50 per cent.

4. The Type of Fractionation to be Used

All separation procedures involve cutting down the complexity of a mixture into discrete groups of sample in a process known as fractionation. This can be through a liquid–liquid extraction, column chromatography, high pressure liquid chromatography (HPLC), etc.

5. The Nature of the Compound

A knowledge about the compound regarding its acidity/basicity/stability/hydrophilicity/lipophilicity would be very helpful. But if the goal of the work is to separate many molecules rather than a single one, the above information would be of little use. In such a case, compounds with extreme properties would be isolated (e.g. nonpolar–polar).

6. The Localization of the Desired Activity

Each potential source of a natural product—whether it is plant, tree, moss, bacteria, vertebrate, invertebrate, insect, terrestrial, or marine-based—has components or parts in which the desired activity or compound is present in greater concentration.

A variety of different techniques can be used for the isolation and purification of natural-product compounds. These techniques include, but are not limited to, solid phase extraction,[26, 27] high-performance liquid chromatography (HPLC), gradient high-performance liquid chromatography,[26, 27] bioautography, thin-layer chromatography (TLC), countercurrent chromatography,[28, 29, 30] droplet countercurrent chromatography,[31] vacuum column chromatography , desalting,[32] liquid–liquid chromatography, paper chromatography,[26, 27] ion-exchange chromatography,[26, 27] size exclusion chromatography,[33] affinity chromatography,[26, 27] acid–base switching technology,[34] centrifugal partition chromatography,[34] liquid–solid chromatography,[26, 27] microwave-assisted extraction,[35] pressurized solvent extraction,[35] large-scale solvent extraction,[26] and supercritical fluid extraction.[36, 37]

GENERAL PROTOCOL FOR ISOLATING PHYTOCHEMICALS

- Identify as to what type of compound, one needs to isolate.
- Choose the solvent for extraction.
- Extract it using appropriate technique.
- Isolate the compound from the extract.
- Characterize it using different analytical techniques.

The steps involved in isolating phytochemicals are shown in Figure 1.10.

Choice of Solvents for Extraction

The selection of solvents depends on the type of phytochemical we are looking for. Organic solvents normally used for extraction are:

n-hexane/benzene
ethylacetate
chloroform — Increasing polarity ↓
methanol
ethanol
water
acetic acid

The extraction can be done based on the polarity of the compound to be isolated or their acidic or basic nature (e.g., phenolics, alkaloids). Polarity arises because of dipole moment. For example, let us consider carbon–carbon bond and carbon–oxygen bond. Oxygen being electronegative, the bond is polarized as $^{\delta+}C—O^{\delta-}$. This implies that the two electrons involved in the bond are more "drawn" towards oxygen thereby creating a very infinitesimal electron deficiency on the carbon. This is the basic reason for polarity. If a chemical does not contain any electronegative atom or in other words no polarized bonds, then it would be nonpolar in nature. For example, *n*-hexane is a solvent obtained during fractionation of petrol and consists of only carbon and hydrogen (hydrocarbon) with no polarized bond and hence is a nonpolar solvent. So too is benzene.

Hence if, one were to extract lipids or fatty acids from a leaf, immersing the leaf in hexane would accomplish a process known as "defatting".

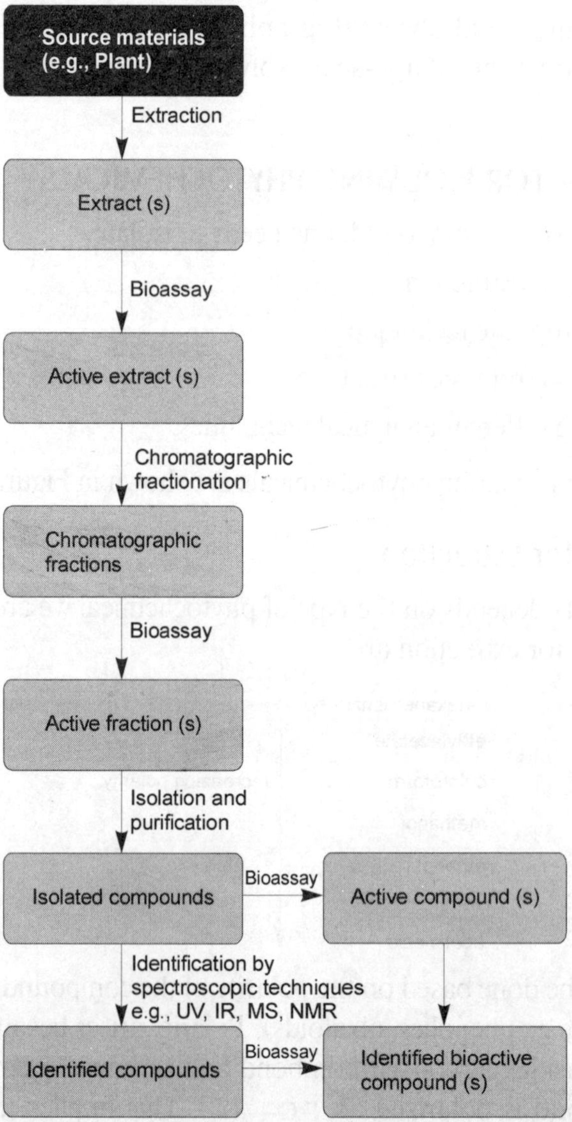

Figure 1.10 Steps involved in isolating phytochemicals

Table 1.8 Variations in chromatographic techniques

Stationary phase	Mobile phase	Name	Utility	
Silical gel coated on a glass slide/alumina sheets	Organic solvents such as hexane, chloroform, methanol, etc.	Thin layer chromatography (TLC)	Qualitative identification, can be used for qualitative separation when used as preparative thin layer chromatography (PTLC)	
Silica gel/alumina packed in a glass column	Organic solvents such as hexane, chloroform, methanol, etc.	Column chromatography	Quantitative separation	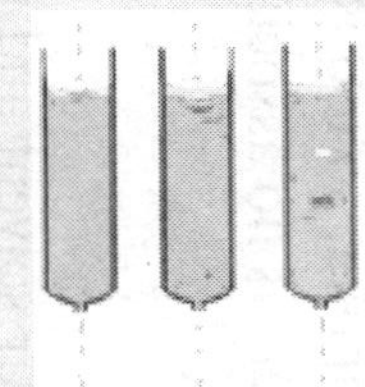
Silical gel coated on alumina sheets	Organic solvents such as hexane, chloroform, methanol, etc.	High performance thin-layer chromatography (HPTLC)	Qualitative identification (Fingerprinting a sample)	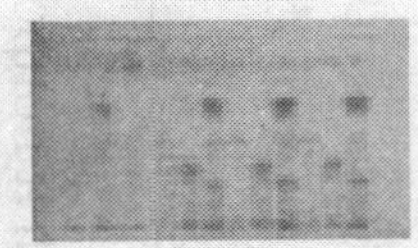
–	–	–	–	

The concept of extraction is the principle of "like dissolves like". A common observation is that any oily matter or grease can be washed easily with kerosene, whereas water alone is not useful. This is because oil/grease is nonpolar, so too is kerosene, hence they associate well whereas water is highly polar, hence there is no association.

Extraction of an Alkaloid

The material is first defatted with a nonpolar solvent and then extracted using an alcohol (methanol or ethanol).This extract is then partitioned between a dilute solution of aqueous tartaric acid and ethyl acetate. The organic layer would contain weakly basic and neutral alkaloids, whereas the aqueous layer would contain basic alkaloids which are neutralized with a weak base (ammonium hydroxide or aqueous sodium carbonate solution). After neutralization, the compound is extracted into an organic phase such as ethyl acetate which is then evaporated to yield a mixture of basic alkaloids, which needs further purification.

But certain alkaloids like cocaine, atropine, etc. can be isolated from the alcholic extract by column chromatography.

Isolation involves the application of chromatographic techniques with all its variation. This includes preparative thin layer chromatography (PTLC), column chromatography, high performance liquid chromatography (HPLC), etc.

For characterization of the compounds spectroscopic techniques such as infrared spectroscopy, nuclear magnetic resonance spectroscopy, mass spectrometry and single crystal X-ray diffraction technique are widely applied.

Extraction of Phenolic Compounds

Polyphenols can be classified into two major groups: 1) phenolic acids (PA) and 2) flavonoids. Because phenolic acids exist in multiple forms, the polarity of each PA can vary significantly. This has lead to difficulty in developing a uniform extraction method for different phenolic acids from varying matrices.

STRUCTURE AND PHYSIOCHEMICAL PROPERTIES OF SOME NATURAL PRODUCTS

Properties such as solubility, acidity, basicity, etc. of any compound is determined by its composition. Table 1.9 gives some of the physiochemical properties associated with the chemical structure of some secondary metabolites.

Table 1.9 Physiochemical properties of some natural products

Classes, some chemical skeletons and examples		General properties
1. Terpenoids		
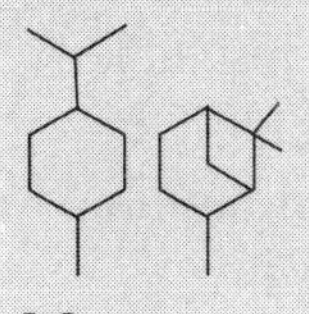 Monoterpenes (C_{10}) e.g., Geraniol, menthol	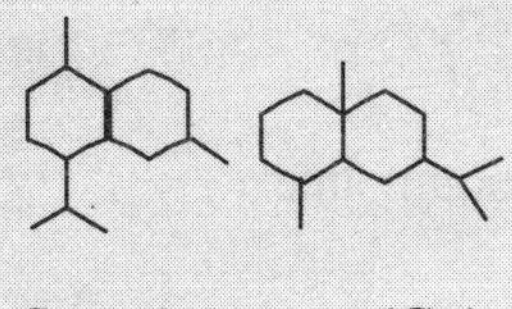 Sesquiterpenes (C_{11}) e.g., Furnesol, α-Cadinene	❁ Water-insoluble ❁ Volatile
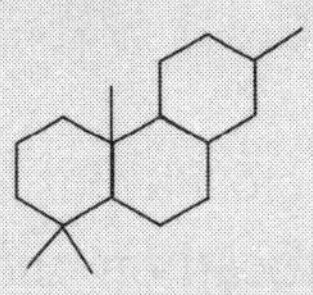	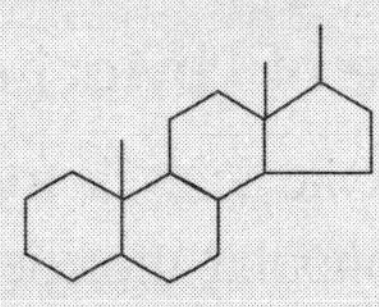	❁ Water-insoluble ❁ Mostly non-volatile
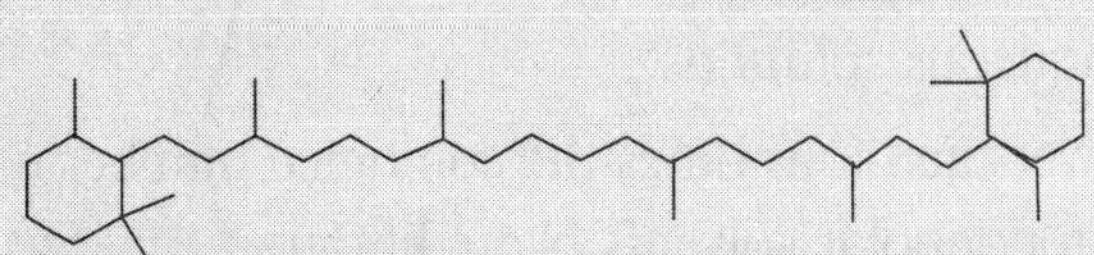 Tetraterpenes (C_{10}), e.g., β-Carotene, lycopene		
2. Phenolics		
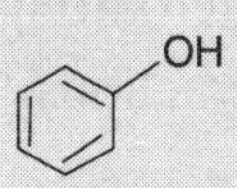 Simple phenols e.g., Vanillin, salicylic acid	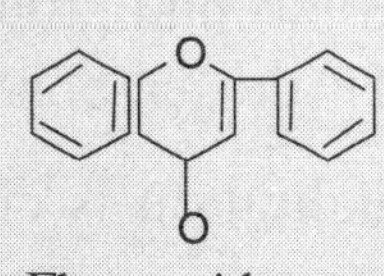 Flavonoids e.g., Catechin, quercetin	❁ Mostly water-insoluble ❁ Oxidized in air ❁ Acidic (anionic) ❁ React with bases to form water-soluble Na and K salts
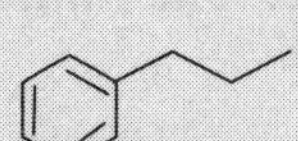 Phenylpropanoids e.g., Caffeic acid, eugenol	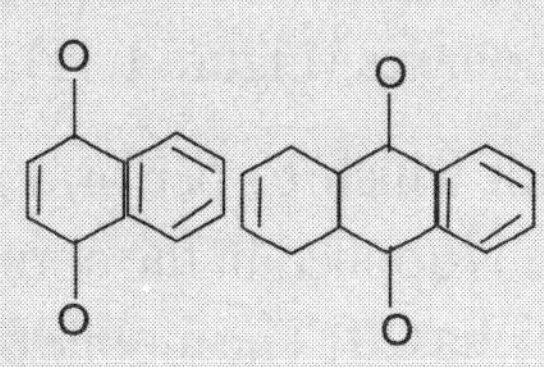 e.g., Quinones Alkannin, alizarin	❁ Mostly water-insoluble ❁ Oxidized in air ❁ Acidic (anionic) ❁ React with bases to form water-soluble salts

(Contd.)

Table 1.9 (Continued)

Classes, some chemical skeletons and examples	General properties
3. Alkaloids 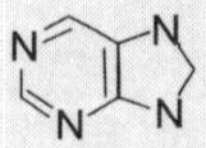 Purine e.g., caffeine, nicotine  Pyridine 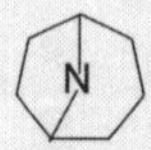 Tropane 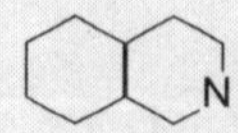 Isoquinoline	❁ Slightly water-insoluble ❁ Basic (cationic) ❁ React with acids to form water-soluble salts ❁ Low MW ones are volatile

Composition that Determines the Nature of Phytochemicals

1. The presence of polar groups such as carboxylic acids, ketones, hydroxyls, etc. increases the water solubility and the presence of nonpolar moieties such as hydrocarbon and halogen decreases the solubility in aqua (aqueous solutions).
2. Compounds that contain carbon atoms fewer than four times the combined number of oxygens and nitrogens are poorly soluble in nonpolar solvents like *n*-hexane.[38]
3. Compounds that contain several hydroxyl amino or carboxylic acid groups are commonly non-volatile.[38]
4. Compounds with aromatic rings, double bonds and groups containing heteroatoms such as O, N, S and P are generally capable of hydrogen-bonding and polar interactions.[39]

One of the widely used technique for qualitation is chromatography. In all the chromatographic techniques, there are two common constants, namely, the stationary phase and the mobile phase. The stationary phase can be silical gel, alumina, octadecyl silane, etc., and the mobile phases can be various organic solvents, aqueous solvents, etc. The name varies based on the stationary and mobile phases (Figure 1.11).

TLC is also the easiest technique with which to perform multidimensional (i.e., two-dimensional) separations. A single sample is applied in the corner of a plate, and the layer is developed in the first direction with mobile phase 1. The mobile phase is dried by evaporation, and the plate is then developed with mobile phase 2 at a right angle (perpendicular or orthogonal direction); mobile phase 2 has different selectivity characteristics when compared with mobile phase 1. In this way, complete separation can be achieved of very complex mixtures (e.g., of the components of a plant extract) over the entire layer surface.

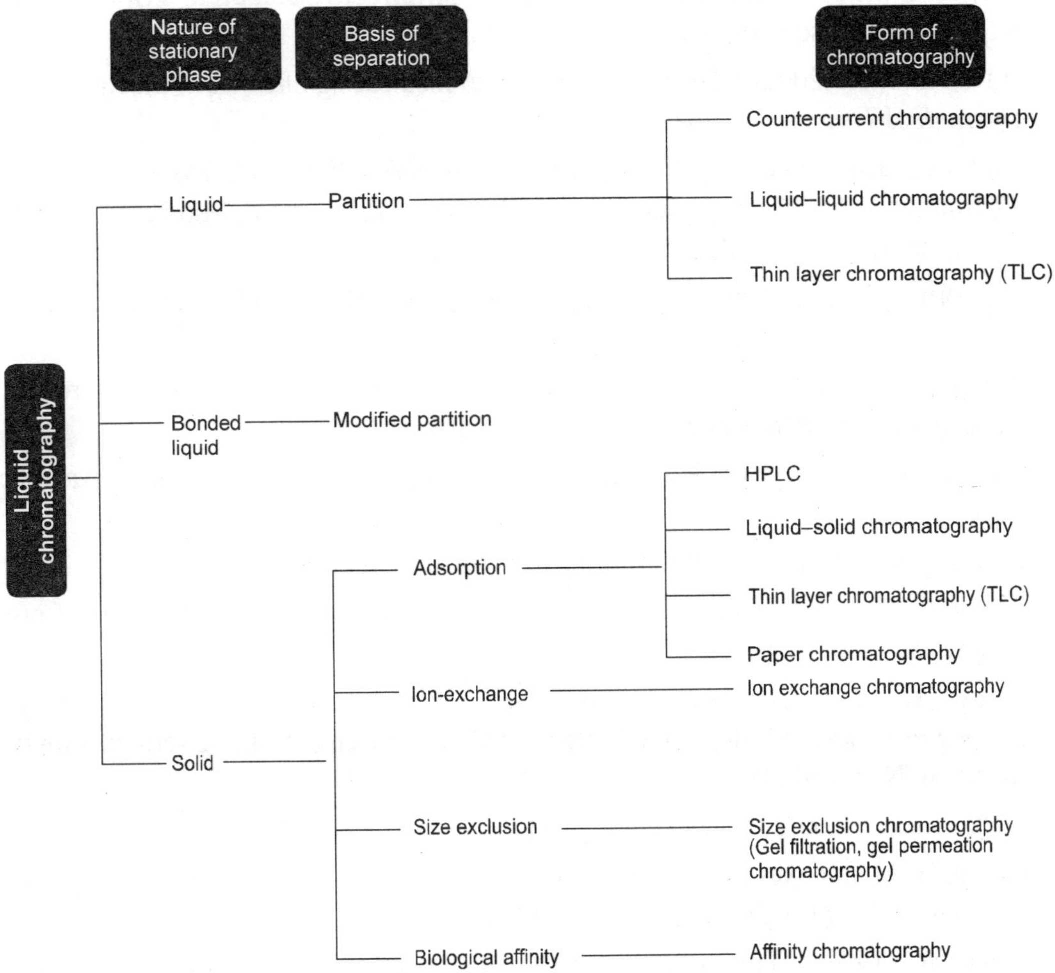

Figure 1.11 Classification of liquid chromatography systems

REFERENCES

1. Holt, G. A., Chandra, A. (2002). "Herbs in the modern healthcare environment—An overview of uses, legalities, and the role of the healthcare professional." *Clin.Res. Regulatory Affairs (USA).* 19, 83–107.
2. Andersen, .M. and Markham, K.R. (Eds.). (2005). *Flavonoids: Chemistry, Biochemistry and Applications.* CRC Press: Boca Raton,
3. Grotewold, E. (Ed.). (2006). *The Science of Flavonoids,* Springer: New York.

4. *USDA Database for the Flavonoid Content of Selected Foods*,(2003). http://www.nal.usda.gov/fnic/foodcomp/Data/Flav/flav.html
5. Havsteen, B.H. (2002). "The biochemistry and medical significance of the flavonoids." *Pharmacol Ther*. *96*: *67*-202.
6. Harborne, J.B. (1986). In: *Plant Flavonoids in Biology and Medicine: Biochemical, Pharmacological, and Structure-Activity Relationships.* Cody, V., Middleton, E.Jr. and Harborne, J.B. (Eds.). Alan R. Liss, New York. pp. 15–24.
7. Harborne, J.B. and Williams, C.A. (2000). "Advances in flavonoid research since 1992." *Phytochemistry. 55*, 481.
8. Treutter, D. (2006). "Significance of flavonoids in plant resistance: A review." *Environmental Chemistry Letters.* 4(3), 147–157.
9. Williams, C.A. and Grayer, R.J. (2004). "Anthocyanins and other flavonoids." *Nat. Prod. Rep. 21*, *539*-573
10. Pawlik, J.R. (1993). Chem.Rev. Marine invertebrate chemical defenses. 93, 1911–1922.
11. Cragg, G.M. and Newman, D. J. (2005). "Biodiversity: a continuing source of novel drug leads. "*Pure Appl. Chem. 77*, 7–24.
12. Goder, K. (1985).*Zur Einführung synthetischer Schlafmittel in die Medizin im 19. Jahrhundert,* Marburger Schriften zur Medizingeschichte; Lang Verlag: Frankfurt/ M., Bern, New York, Vol. 18.
13. Phillipson, J.D. (2001). "Phytochemistry and medicinal plants." *Phytochemistry. 56*, 237–243.
14. Cragg, G. M. and Newman, D. J.(2001). "Natural product drug discovery in the next millennium." *Pharmaceutical Biology*, 39, S8–17.
15. Noble, R. (1990). "The discovery of the vinca alkaloids–Chemotherapeutic agents against cancer." *Biochem. Cell Biol.* 68, 1344–1351.
16. Cragg, G.M. (2002). "Natural product drug discovery and development: The United States National Cancer Institute Role." *Puerto Rico Health Sciences Journal.* 21, 97–111.
17. Newman, D.J., Crag, G.M. and Snader, K.M. (2003). "Natural Products as Sources of new drugs over the Period 1981–2002." *J. Nat. Prod.* 2003, 66, 1022–1037.
18. Farnsworth, N.R., Akerele, O., Bingel, A.S., Soejarto, D.D. and Guo, Z. (1985). "Medicinal plants in therapy." *Bull WHO*, 63, 965–981.
19. Grabley, S. and Sattler, I. (2003). "Natural products for lead identification: Nature is a valuable resource for providing tools." In: Hillisch,A. and Hingenfeld,R. (Eds.). *Modern Methods of Drug Discovery.* Birkhäuser Verlag, Switzerland, pp. 87–107.

20. Dahanukar, S. A., Kulkarni, R. A. and Rege, N.N. (2000). "Pharmacology of medicinal plants and natural products." *Ind. J. Pharmacol.*, 32, S81–S118.

21. Nakanishi, K. (1999). "An historical perspective of natural products chemistry." *Comprehensive Nat. Prod. Chem.* 8, xxi–xxxviii.

22. Nakanishi, K. (1999). "An historical perspective of natural products chemistry." In: Ushio, S. (Ed.). *Comprehensive Natural Products Chemistry*, Vol. 1. Elsevier Science B.V., Amsterdam. pp. 23–40.

23. DerMarderosian, A. and Beutler, J. A. (Eds.), *The Review of Natural Products*, 3rd edn. Facts and Comparisons, St. Louis, MO. pp. 824–826.

24. Meyer-Warnod, B. (1984). "Natural essential oils: extraction processes and applications to some major oils." *Perf. Flav.* 9, 93–103.

25. Klesper, E. (1980). "Chromatography with supercritical fluids." In: *Extraction with Supercritical Gases.* Schneider, G.M., Stahl, E. and Wilke, G. (Eds.). Verlag Chemie, Weinheim, Germany.

26. Olmedo, D., Rodriguez, N., Vasquez, Solis, P.N., Lopez-Pe', J.L., San Feliciano A. and Gupta, M.P. (2007). "A new coumarin from the fruits of *Coutarea hexandra.*" Natural Product Research. 21(7) , 625–631.

27. Cannell, R.J.P. (1998). "Chem inform abstract: follow-up of natural product isolation." *Cheminform*, 30: doi: 10. 1002/Chin. 19990128.

28. Cannell, R.J.P. (1998). "How to approach the isolation of a natural product." In: Cannell, R.J.P. (Ed.). *Methods in Biotechnology*, Vol. 4: *Natural Products Isolation.* Humana Press Inc., Totowa, NJ. pp. 1–51.

29. Alvi, K.A. (2000). "A strategy for rapid identification of novel therapeutic leads from natural products." In: Cutler, S. and Cutler, H. (Eds.). *Biologically Active Natural Products.* CRC, Boca Raton, FL. pp. 185–195.

30. Foucault, A.P. and Chevolot, L. (1998). "Counter-current chromatography: Instrumentation, solvent selection and some recent applications to natural product purification." *J. Chromatogr. A* 808(1–2), 3–22.

31. Foucault, A.P. (1997). "Recent advances in purification of natural products by countercurrent chromatography." In: Schreier, P. (Ed.). *Natural Products Analysis: Chromatography, Spectroscopy, Biological Testing*, Symposium, Wuerzburg, Germany, Vieweg, Wiesbaden, Germany. pp. 13–25.

32. Hostettmann, K., Hamburger, M., Hostettmann, M. and Marston, A. (1991). "New developments in the separation of natural products." *Recent Adv. Phytochem.* (Modern Phytochemical Methods) 25, 1–32.

33. Shimizu, Y. (1998). "Purification of water-soluble natural products." *Methods in Biotechnol., (Natural Products Isolation).* 329–341.

34. Horton, T. (1972). "Large-scale gel filtration for purification of natural products." *Am. Lab.* 4(5), 83–84, 86–88, 90–91.

35. Ellis, N. (2001). "Extracting APIs from natural products." *Speciality Chem. Mag.* 21(6), 10–12.

36. Kaufmann, B. and Christen, P. (2002). "Recent extraction techniques for natural products: Microwave-assisted extraction and pressurized solvent extraction." *Phytochem. Anal.*, 13(2), 105–113.

37. William, K. Modey, Dulcie A. Mulholl and Mark, W. Raynor. "Analytical supercritical fluid extraction of natural products." *Phytochem. Anal.*, Vol. 7(1), 1–15.

38. Puri, R. K. (1998). "Supercritical fluid extraction and its applications in natural products. *Ind. J. Nat. Prod.*, 14(2), 3–23.

39. Van Beek, T.A. (1999). "Modern methods of secondary product isolation and analysis." In: *Chemical from plants :Perspectives on plant secondary products.* Walton, N.J. and Brown, D.E. (Eds.). Imperial College Press , London, p.91

40. Thurman, E.M. and Mills, M.S. (1998). *Solid Phase Extraction :Principles and Practice.* John Wiley and Sons, Inc. New York.

2

ISOLATION TECHNIQUES IN NATURAL PRODUCTS

EVOLUTION OF CHROMATOGRAPHY

The most widely and extensively used technique in isolation of natural products is chromatography. This powerful technique was brought into shape by a Russian botanist M.Tswett in 1906 when he separated chloroplast pigment[1] using calcium carbonate (solid phase) and petroleum ether (liquid phase). During the separation, bands of different colours traversed the column and Tswett coined the term "chromatography" which is derived from the Greek word "*chroma*" meaning "colour" and "*graphia*" meaning "to write", that is, "to write with colours".

No other discovery has exerted as great an influence and widened the field of investigation of the organic chemist as Tswett's chromatographic adsorption analysis.

In 1940, Tiselius earned the Nobel Prize for his pioneering work in adsorption and electrophoresis.[2] Martin and Synge developed a model to describe column efficiency and evolved the liquid–liquid chromatography technique and received the Nobel Prize in 1952. Consden *et al.* discovered the paper chromatography technique in 1944. Cremer in 1951 developed the gas–solid chromatography,[5] while James and Martin contributed to the appearance of gas–liquid chromatography.[6] In 1964, Horvath introduced a chromatography technique that was the reverse of Tswett's invention in terms of its nature[7] and hence termed it as reverse-phased chromatography. Giddings postulated the widely used theories on chromatography in 1965.[7]

In 1963, the first commercial liquid chromatograph, the Model GPC 100 Liquid Chromatograph, was introduced by Waters. In 1967, the Waters ALC 100 HPLC made chromatography a tool that was affordable for most laboratories. The evolution and the use of liquid chromatography from the 1970s until today have been explosive.

WHAT IS CHROMATOGRAPHY

Chromatography is a separation technique wherein separations are based on a forced transport of the mobile phase (liquid, gas, etc.) carrying the analyte mixture over the stationary phase (solid, liquid, etc.).The differences in the interactions of analytes in the mobile phase with the stationary phase result in different migration times for the components in the mixture. In the above definition, the presence of two different phases is stated and consequently there is an interface between them. One of these phases provides the analyte transport and is usually referred to as the mobile phase, and the other phase is immobile and is typically referred to as the stationary phase. A mixture of components, usually called analytes, are dispersed in the mobile phase at the molecular level allowing for their uniform transport and interactions with the mobile and stationary phases. Analyte molecules undergo multiple phase transitions between the mobile phase and adsorbent surface. Average residence time of the molecule on the stationary phase surface is dependent on the interaction energy. For different molecules with very small interaction energy difference, the presence of significant surface is critical, since the higher the number of phase transitions that analyte molecules undergo while moving through the chromatographic column, the higher the difference in their retention. The nature of the stationary and the mobile phases, together with the mode of transport through the column, is the basis for the classification of chromatographic methods. In all these techniques, the common denomination are the stationary phase and the mobile phase and what varies is the nature of these phases (Figure 2.1).

Thus the resolution (extent of separation) of species by chromatography depends on the triad of interactions (mobile phase/solute/stationary phase) for each solute. All the information from a chromatographic experiment are recorded in the chromatogram. This includes the minimum number of sample components (peak count), identification of separated components (peak location), composition (peak areas), system properties (peak width and peak shape), and sample properties (peak location and width).

CHROMATOGRAPHY AND ITS METHODOLOGICAL CLASSIFICATION

Chromatographic methods may be classified as follows:[8]

- Thin layer chromatography (High performance, preparative)
- Paper chromatography

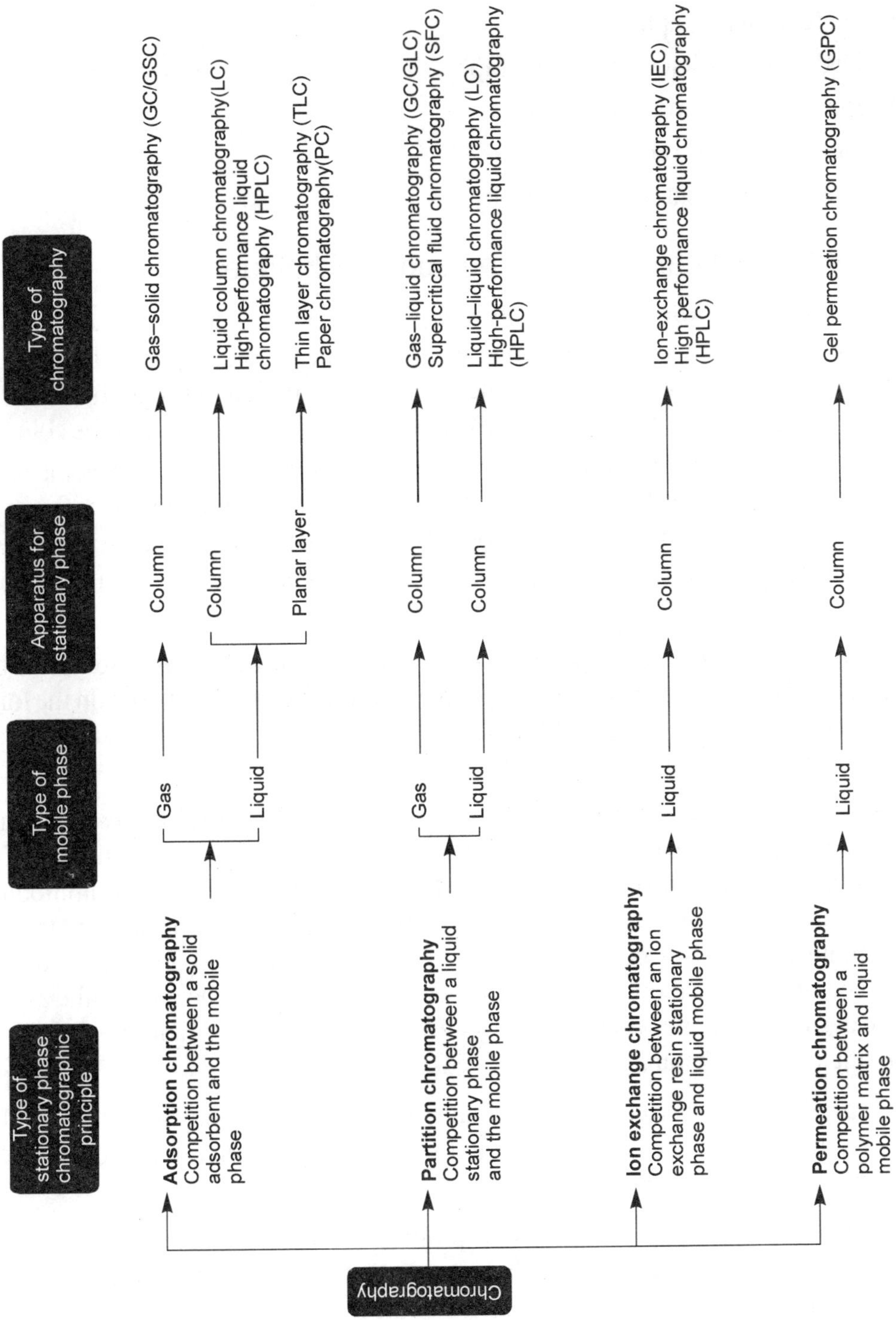

Figure 2.1 Classification of chromatgraphic methods

- Column chromatography
- Gas chromatography
- High-pressure liquid chromatography
- Ion-exchange chromatography
- Gel filtration chromatography
- Supercritical fluid chromatography

Chemical Forces behind Chromatographic Techniques

1. *Gel permeation, or size exclusion chromatography* The mechanism of separation based on diffusive accessibility of stationary phase pore volume to solute species. Large molecules are excluded from the pores of the packing, thus passing rapidly through the column in the interstitial volume. Small molecules can enter packing pores, and thus exit the column later.

2. *Hydrophobic RP-HPLC (reverse phase chromatography)* The hydrophobic surface of these packings interacts with the hydrophobic regions of the solute molecule. Separation of solutes is based on differences in the hydrophobicity of the molecules. There is an interaction of nonpolar region of protein with nonpolar stationary phase. The selected mobile phase has effect of displacing this interaction.

3. *Ion-exchange chromatography* The chromatography medium possesses ionizable sites on its surface. These sites can be either anionic or cationic. Retention is based on the formation of ion pairs between the solute and the packing. This in turn is conditioned by the pH and ionic strength of the mobile phase.

4. *Affinity chromatography* Affinity chromatography utilizes the affinity of a solute in a mixture, towards a chemical immobilized on the stationary phase. If there is mixture of protein with a particular protein having affinity towards, say, Ni^{2+} metal, the metal is immobilized on the stationary phase and when the solution of protein is passed over it, only that protein having affinity towards Ni^{2+} gets bound while the remaining are eluted. Then the bound protein can be separately washed out by varying solvent or pH. This can be a general class process (triazine dye l and with nucleotide binding sites), or highly specific (a monoclonal antibody ligand with its corresponding antigen).

5. *Normal phase chromatography* Normal phase chromatography utilizes a highly polar stationary phase and nonpolar (nonaqueous) solvents. Separation is based on differences in polar regions of solute molecules.

6. *Chiral chromatography* This type exploits the energy differences and differential interaction of stereoisomeric molecular sites with ligands. Such ligands can either be immobilized on the packing surface or carried in the mobile phase.

THIN-LAYER CHROMATOGRAPHY (TLC)

Thin-layer chromatography serves as one of the many methods in providing a chromatographic plant extract fingerprint,[9] i.e., every plant extract C methanolic, hexane etc. is very characteristic of the plant (through the percentage of components may vary for various geography). For example, the ethylacetate extract of the leaves of Tulsi plant should yield a similar chromatographic pattern under similar elution conditions.

Thin-layer chromatography (TLC) consists of the sample solution being applied as a spot or band on the origin of a layer spread on a support (the plate). After evaporation of the sample solvent, the plate is placed in a sealed chamber or tank that contains a solvent mixture chosen as the mobile phase. Development occurs as the mobile phase moves up the layer by capillary forces. Instrumental development methods, such as over-pressured layer chromatography (OPLC) or automated multiple development (AMD), can provide separations with increased resolution.[10, 11] The plate is removed from the chamber, and the separated zones are detected by physical or chemical methods, identified by comparison of their R_f values (R_f = distance of migration of the sample zone/distance of the mobile phase front) and colours to standard zones on the same plate, and quantified by visual or instrumental densitometry based on measurement of zone sizes and intensities. Zone identification is confirmed by off- or on-line coupling of TLC with visible/ultraviolet (UV), Fourier transform infrared (FTIR), Raman, and mass spectrometry (MS).

TLC is a mode of liquid chromatography in which the sample is applied as a small spot or streak to the origin of a thin sorbent layer such as silica gel, alumina, cellulose powder, polyamides, ion exchangers or chemically bonded silica gel supported on a glass, plastic, or metal plate. This layer consists of finely divided particles and constitutes the stationary phase. The eluent or mobile phase is a solvent or a mixture of organic and/or aqueous solvents in which the spotted plate is placed. The mobile phase moves through the stationary phase by capillary action, sometimes assisted by gravity or pressure.

Plates can be visualized, depending on the chemical structure of the compounds at visible light, UV (254 nm and 365 nm) or by using spray reagents. The effectiveness of the separation depends on the mixture to be separated, the choice of the mobile phase and the adsorption layer.[12]

Thin-layer chromatography (TLC) is one of the most popular and widely used separation techniques because of its ease of use, cost-effectiveness, high sensitivity and speed of separation, as well as its capacity to analyse multiple samples simultaneously. It has been applied to many disciplines including biochemistry[13, 14] toxicology,[15, 16] pharmacology,[17, 18] environmental science,[19] food science,[20, 21] and chemistry.[22, 23]. TLC can be used for separation, isolation, identification, and quantification of components in a mixture. It can also be utilized on the preparative scale to isolate an individual component.

A thin-layer of silica gel/cellulose/alumina/polyamide is coated over a glass slide/aluminium or plastic sheet (pre-coated plates) of roughly 25 mm thickness. This thickness would suffice for identification of the components and if separation is the aim then a thicker coating on a bigger plate is applied. A dilute solution of the mixture to be resolved is applied (spotted) using a capillary tube.

The plate is developed in a closed chamber that is generally saturated with the developing solvent(s). The sample is dissolved in an appropriate solvent and applied as spots or bands along one side of the sorbent layer, approximately 1 cm from the edge. An eluent (single solvent or solvent mixture) is allowed to flow by capillary action through the sorbent starting at a point just below the applied samples. Most commonly this is achieved by using a glass rectangular tank in which the eluent is poured to give a depth of about 5 mm. The plate is placed in the tank or chromatography chamber and the whole is covered with a lid. As the eluent front migrates through the sorbent, the components of the sample also migrate, but at different rates, resulting in separation. When the solvent front has reached a point near the top of the sorbent layer, the plate or sheet is removed and dried. The spots or bands on the developed layer are visualised, if required, under UV light or by chemical treatment or derivatization. For quantitative determinations, zones can be removed or eluted from the layer, or the plate can be scanned at pre-determined wavelengths without disturbing the layer surface. The modern use of TLC has seen a strong move in the direction of plate scanning and video imaging as a means of providing sensitive and reliably accurate results and a more permanent record of the chromatogram. This is in addition to its obvious labour-saving aspect and chemically "clean" approach.

Silica gel is polar, hence the polar compounds in the mixture interact with the gel through weaker hydrogen bonds and if the mobile phase is nonpolar like hexane, the nonpolar compounds move well with this phase and are eluted at the top of the plate. Figure 2.2 depicts the resloution concepts in TLC.

Plant chemistry research applies chromatography as a principal separation technique. It can be used in a search for the optimum extraction solvents, for identification of known and unknown compounds, and—what is at least equally important—for selection of biologically active compounds. TLC also plays a key role in preparative isolation of compounds, purification of the crude extracts, and control of the separation efficiency of the different chromatographic techniques and systems. TLC has many advantages in plant chemistry research and development. These include single use of stationary phase, wide optimization possibilities with the chromatographic systems, special development modes and detection methods, storage function of chromatographic plates (all zones can be detected in every chromatogram by multiple methods), low cost in routine analysis, and availability of purification and isolation procedures.

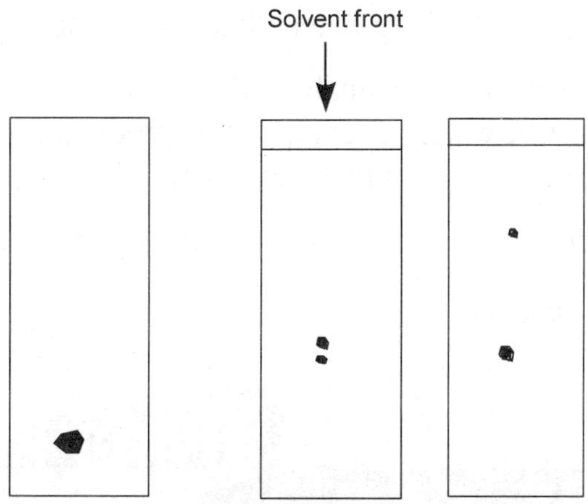

Figure 2.2 A schematic representation of resolution in TLC

Prerequisites for a Compound to be Amenable to TLC Analysis

- Should be soluble in a solvent or mixture of solvents. This is essential to load the TLC plate.
- Should have low volatility, because highly volatile compounds do not stay long enough to interact with stationary/mobile phases.
- Should be neither too polar nor too non-polar. This has been quantified based on the stationary phases like silica gel, alumina and mobile phases from benzene to methanol. Water or its presence deactivates the silica gel.

Application Area of Thin-Layer Chromatography

Apart from plant chemistry, the other areas of application of thin-layer chromatography include the following.

Clinical chemistry, forensic chemistry and biochemistry Determination of active substances and their metabolites in biological matrices, diagnosis of metabolic disorders such as PKU (phenylketonuria), cystinuria and maple syrup disease in babies.

Cosmetology Dye raw materials and end products, preservatives, surfactants, fatty acids, constituents of perfumes—to detect the quality.

Food analysis Determination of pesticides and fungicides in drinking water, residues in vegetables, salads and meat, vitamins in soft drinks and margarine, banned additives (e.g., sandalwood extract in fish and meat products), compliance with limit values (e.g., polycyclic compounds in drinking water, aflatoxins in milk and milk products).

Environmental analysis Groundwater analysis, determination of pollutants from abandoned armaments in soils and surface waters, decomposition products from azo dyes used in textiles. Figure 2.3 represents the percentage of the areas where TLC is applied.

Analysis of inorganic substances Determination of inorganic ions (metals).

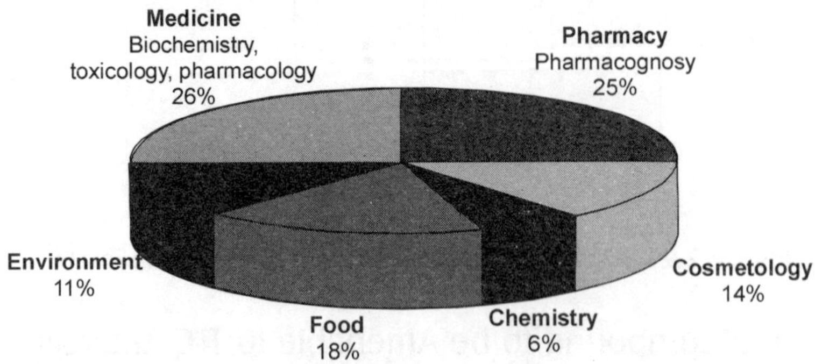

Figure 2.3 Fields of application of thin-layer chromatography (TLC/HPTLC) over the years

The Indicator in TLC—Retardation Factor

Each metabolite/compound has a characteristic retardation factor(R_f). This can be defined as the distance of the component from the origin/distance of solvent front from origin. (Figure 2.4).

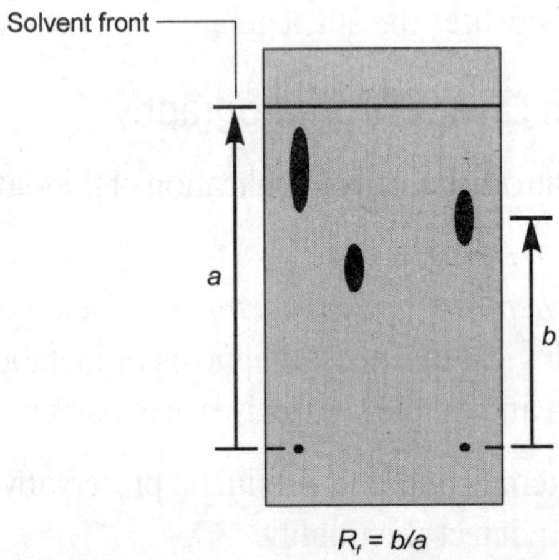

Figure 2.4 Calculation of retardation factor

Protocol for Running a Thin-Layer Chromatographic Analysis

1. Pre-coated silica gel plates can be cut into 6 × 3 cm. If the plates are prepared over glass slides, they need to be activated before spotting. It is done by heating it in an oven to ~150°C.
2. Dissolve the extract/mixture in a suitable solvent and spot a very dilute solution using a capillary tube.
3. Spot the plate using an applicator or a capillary tube.

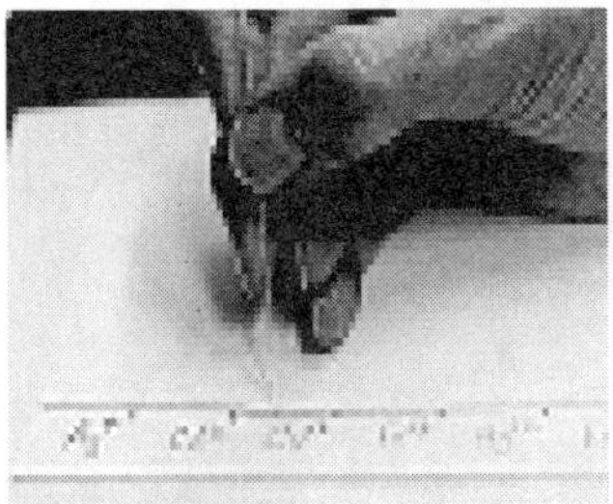

4. After spotting, the plate is placed inside a developing chamber with the mobile phase in it.

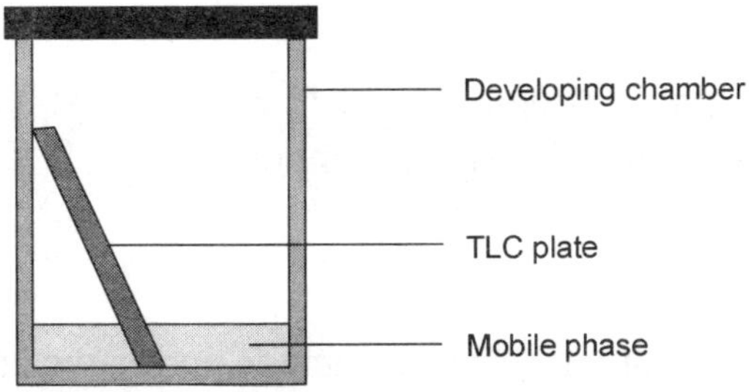

Care should be taken that the spotted area is not immersed in the solvent.

5. The mobile phase which can be a single solvent (e.g., ethyl acetate) or mixture of solvents (ethyl acetate : hexane, 4 : 6 ratio) is chosen (by trial and error) so that the R_f value lies in the range of 0.4–0.7. The plate is developed in the solvent wherein the solvent rises by capillary action against gravity. Only organic solvents like hexane, ethyl acetate, chloroform, methanol/ethanol or their combinations are used in thin-layer chromatographic analysis. Water and water-containing solvents are to be avoided as they deactivate the silica gel resulting in no-separation.
6. The mixture is carried by the solvent and depending on the nature of its constituents, there is a preferential affinity either towards the stationary phase or the mobile phase. For example, a nonpolar compound has less affinity towards the polar silca gel (stationary phase) and moves faster than a polar compound. The migration of each component in a mixture during TLC is a result of two opposing forces: capillary action of the mobile phase

and retardation action of the stationary phase. Both forces contribute to achieve differential migration of each component.

7. The TLC plate is taken out of the chamber and air/oven-dried.

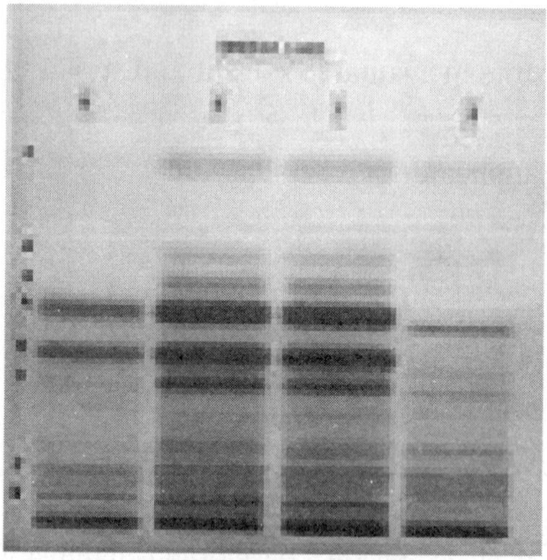

Figure 2.5 A chromatographic plate with multipe compounds spotted

The chromatographic plate, (Figure 2.5) depicts well-resolved spots in mixtures of compound. Each spot corresponds to a single compound.

Other Features of TLC

Types of stationary phase that are widely used include, inorganic adsorbent layers (silica or silica gel and alumina); organic layers (polyamide, cellulose); organic, polar, covalently bonded modifications of the silica gel matrix (diol, cyanopropyl, and aminopropyl); and organic, nonpolar, bonded stationary.

Sorbents applied in TLC have different surface characteristics and, hence, different physicochemical properties.

Multiple samples can be analysed at the same time on a single TLC reducing the time and solvent volume used per sample.

There are special modes of development in TLC:[24]

Multiple development consists in repeated development of the same plate with mobile phase over the same distance and in the same direction uni-dimensional multiple development (UMD). UMD consists of repeated development of the chromatogram over the same development distance, with a given mobile phase of constant composition and with drying the plate between the individual development runs. After each development, the plates are dried in air. Each consecutive development results in band re-concentration and thus increases the efficiency of the separation.

Incremental multiple development (IMD), IMD is performed by the stepwise increase in the development distance (the increment in the development distance is kept constant), using a steady mobile phase composition and drying the plate between the development runs. It results in narrowing of the spots or zones and improved resolution. In incremental multiple development (IMD) the development distance is increased linearly in 10 or 20 mm steps with evaporation of the mobile phase from the plates after each step, using the same solvent or, for the modified IMD technique (bivariant multiple development).

In gradient multiple development (GMD), each step of the repeated chromatogram development is performed with a mobile phase of different composition, thus enabling gradient development. The development distance of the consecutive development runs is kept steady and it is only the mobile phase composition that changes, thus enabling the analysis of complex mixtures spanning a wide polarity range. Separation of the low-polarity components is achieved on a silica layer. When a medium polarity mobile phase is used, the medium polarity components are separated (the first group is then eluted to the upper edge of the plate). With the high-polarity mobile phases, separation of the high-polar components of plant extracts can be obtained.

In bivariant multiple development (BMD) a stepwise change both of the development distance and the mobile phase composition is effected using a special chamber, and a computer program, an improved version, known as automated multiple development (AMD) can be applied, with the distance of the development increasing and the mobile phase strength decreasing at each step. AMD enables the analysis of complex samples over a wide polarity range and provides focusing (tightening) of the zones. In the circular and anticircular development modes, the mobile phase migrates radially from the centre to the periphery or from the periphery to the centre, respectively. Analytes with lower RF values are better resolved by means of circular chromatography than by means of linear chromatography, and the advantage of the anticircular mode is that it allows better resolution of compounds with higher RF values

TLC is also the easiest technique with which to perform multidimensional (i.e., two-dimensional) separations. A single sample is applied in the corner of a plate, and the layer is developed in the first direction with mobile phase 1. The mobile phase is dried by evaporation, and the plate is then developed with mobile phase 2 at right angles (perpendicular or orthogonal direction);

Name of the reagent	Composition	Detected compounds
Neu-reagent	1% 2-amino ethyl diphenyl borinate dissolved in methanol or ethanol	Compounds possessing hydroxyl groups, such as flavonoids, sugars, anthocyaninis, or hydroxy acids.
van Urks reagent	50 mg 4-dimethylaminobenzaldehyde with 1 mL concentrated H_2SO_4 dissolved in 100 mL ethanol	Terpenoids, sterols and alkaloids—lipophilic compounds—forming dark-coloured zones
Vanillin in sulphuric acid	(i) 1% ethanolic vanillin solution (ii) 10% ethanolic sulphuric acid Dip or spray the plate with (a) followed immediately by (b) Evaluate the plate in visible light. after heating at 110°C for 5–10 min. For dipping, 0.4 g vanillin are dissolved in 100 mL diethyl ether and finally , 0.5 mL concentrated sulphuric acid are added. Heat the plate after dipping for 3–5 min. at 110°C and evaluate in visible light.	Essential oils
Ehrlich's reagent	Dissolve 0.3 g of 4-dimethylaminobenzaldehyde in 25 mL methanol. Add 10 mL 32% HCl while cooling. The addition of one drop of a 10% aqueous iron-II-chloride solution mostly gives improved results. The plate can be sprayed or dipped.	Iridoids and proazulenes
Eckert's reagent	Anisaldehyde (4-methoxybenzaldehyde) in combination with sulphuric acid, which reacts with sugars and glycosides.	Sugars, glycosides
Marquis reagent	Dilute 3 mL formaldehyde to 100 mL with concentrated sulphuric acid. Evaluate the plate in visible light immediately after spraying or dipping.	Opiates—morphine, codeine, thebaine
Gibbs reagent	Dissolve 50 mg of 2,6-dichloroquinone chloroimide in 200 mL ethyl acetate or methanol. Spray or dip the plate and immediately expose to ammonia vapour.	Sulphur-containing compounds, arbutin and capsaicin

(Contd.)

Name of the reagent	Composition	Detected compounds
Carr–Price reagent	4% $SbCl_3$ in 50 mL $CHCl_3$ or 50 mL glacial acid.	Forms various colours, characteristic for compounds showing carbon double bonds
Ninhydrin (1,2,3,-Indantrione)	Dissolve 30 mg ninhydrin in 10 mL *n*-butanol and mix 0.3 mL 98% acetic acid. Dip or spray and heat for 5–10 min. under observation and evaluate in visible light.	All samples containing $—NH_2$ groups such as amino acids, peptides, or amines into red or purple products. To perform a reaction at least 5–10 min. heating at 120°C is necessary.
DNPH reagent	Dissolve 0.1 g of 2,4-dinitrophenyl-hydrazine in 100 mL methanol, followed by the addition of 1 mL of 36% hydrochloric acid. Spray or dip the plate and evaluate immediately in visible light. Heating may be needed.	Aldehydes and ketones
Liebermann–Burchard reagent	1 mL of concentrated H_2SO_4 and 20 mL acetic anhydride to 100 mL with chloroform (or absolute ethanol).	Sterols and terpenoides
Barton's reagent	(i) 1 g potassium hexacyanoferrate (III) dissolved in 100 mL water. (ii) 2 g iron-III-chloride in 100 mL water. Spray or dip the TLC plate in a 1:1 mixture of (i) and (ii).	Gingeroles
Dragendorff reagent	*Solution A* Dissolve 0.85 g bismuth nitrate in 10 mL glacial acetic acid and 40 mL water under heating. If necessary, filter. *Solution B* Dissolve 8 g potassium iodide in 30 mL water. Just before spraying or dipping, solution A and B are mixed with 4 mL acetic acid and 20 mL water.	Alkaloids and heterocyclic nitrogen compounds
Iron-III-chloride reagent	Dissolve 1 g Iron-(III)-chloride in 5 mL water and dilute to 100 mL with ethanol. Spray or dip the plate and evaluate in visible light.	Phenols, flavonoids, tannins, plant acids, ergot alkaloids.

mobile phase 2 has different selectivity characteristics when compared with mobile phase 1. In this way, complete separation of very complex mixtures (e.g., of the components of a plant extract) can be achieved over the entire layer surface.

London dispersion or van der Waals forces, dipole–dipole interaction, intermolecular adsorption and hydrogen bonding are the main forces that govern in TLC. They can be between stationary and mobile phase, or between molecules themselves.

Visualizing Agents/Methods in Thin-layer Chromatography[25, 26]

1. **Ultraviolet radiation** The developed TLC plate is exposed to UV radiation and if the compounds are UV-sensitive, i.e., possess extended conjugation/chromophores, etc., they would fluoresce.
2. **Iodine vapours** The developed TLC plate in placed in an iodine chamber (with saturated vapours of iodine). Iodine is lipophilic and accumulates in lipophilic sample spots, showing a brown colour on a pale yellow-brown background. The same result will occur by spraying with an iodine solution. In nearly any case this iodine accumulation is totally reversible without altering the sample because outside the closed chamber iodine evaporates quickly from the plate.
3. **Methanolic sulphuric acid** A dilute (10%) solution of sulphuric acid in methanol is sprayed on the developed TLC plate and heated. This results in the charring of spots due to the elemental reduction of carbon by sulphuric acid. This decomposition process of organic samples mainly results in black-to-brown zones on a white background.
4. **Pancol D** 21.6 g of ammonium heptamolybdate + 1.6 g of ceric sulphate in 50 ml of concentrated sulphuric acid is made up to 500 ml specific for terpenoids.

HIGH-PERFORMANCE THIN-LAYER CHROMATOGRAPHY (HPTLC)

High-performance thin-layer chromatography is a sophisticated and automated form of TLC. Whatever can be resolved through TLC, can be quantified with HPTLC.

Features of HPTLC

1. Simultaneous processing of sample and standard—better analytical precision and accuracy and less need for internal standard
2. Lower analysis time and less cost per analysis
3. Low maintenance cost
4. Simple sample preparation, can handle samples of divergent nature
5. No prior treatment for solvents like filtration and degassing

6. Low mobile phase consumption per sample
7. No interference from previous analysis; fresh stationary and mobile phases for each analysis—no contamination
8. Visual detection possible—open system
9. Non-UV-absorbing compounds detected by post-chromatographic derivatization
10. The fingerprints of samples can be obtained and the components can be detected and described without the need to know the chemical nature of each chromatographic spot.
11. The HPTLC fingerprints can be generated and saved as electronic images using video or digital cameras or a flat-bed scanner.
12. In industrial processes based on fingerprints, any change or degradation during process development can be detected; can be documented if the composition of individual batches is kept constant and if the raw material is preserved or converted in the steps of production process.

The major differences between HPTLC and TLC are listed in the following Table.

	HPTLC	TLC
Layer of sorbent	100 μm	250 μm
Efficiency	High due to smaller particle size generated	Less
Separations	3–5 cm	10–15 cm
Analysis time	Shorter migration distance and the analysis time is greatly reduced	Slower
Solid support	Wide choice of stationary phases like silica gel for normal phase and C_8, C_{18} for reversed phase modes	Silica gel, alumina and Kiesulguhr
Development chamber	New type that requires less amount of mobile phase	More amount
Sample spotting	Auto sampler	Manual spotting
Scanning	Use of UV/visible/fluorescence scanner that scans the entire chromatogram qualitatively and quantitatively and the scanner is an advanced type of densitometer	Not possible

Steps Involved in HPTLC

1. Selection of chromatographic layer
2. Sample and standard preparation

3. Layer pre-washing
4. Layer pre-conditioning
5. Application of sample and standard
6. Chromatographic development
7. Detection of spots
8. Scanning
9. Documentation of chromatic plate

Selection of Chromatographic Layer and Running HPTLC

Varied pre-coated plates with different support materials are being used. 80% of the analysis is carried out using silica gel GF. For basic substances like alkaloids and steroids, aluminium oxide is the preferred stationary phase. Cellulose is the stationary phase used for separation of amino acids, dipeptides, sugars and alkaloids. For non-polar substances like fatty acids, carotenoids, cholesterol, etc., reverse phases such as RP2, RP8 and RP18 are used. Plates exposed to high humidity or kept on hand for long time are to be activated in an oven at 110–120°C for 30 minutes prior to spotting. Care should be taken to avoid interference from impurities and water vapour. The commonly used mobile phases include, methanol, chloroform, ethyl acetate, ammonia, methylene chloride and 1% acetic acid. The solvents are selected based on trial and error/ literature.

The concentration of the sample/extract to be evaluated is usually in the range of 0.1–1 μg / μl, above which poor separation results. Unlike TLC, the application of the sample is bandlike. As a preconditioning measure the HPTLC chamber needs to be saturated with the solvent vapours (a filter paper can be used), since unsaturated chamber causes high R_f values. Similar to TLC, HPTLC plates are first visualized under UV for any spots.

Quantification

The sample and standard should be chromatographed on the same plate. After development, the chromatogram is scanned. The HTLC scanner can scan the chromatogram in reflectance or in transmittance mode by absorbance or by fluorescent mode. The scanning speed is selectable up to 100 mm/s and the spectra recording is fast—36 tracks with up to 100 peak windows can be evaluated. The degradation products induced during a stress test could be detected. Stability tests using HPTLC are based on the visual comparison of fingerprints, which are not used for very small changes in the sample to be detected, but significant changes are easily seen. A great advantage of HPTLC is that the entire sample on the plate, which is not possible in on-column separations (gas chromatography and liquid chromatography).

An example is the stability tests of *Vitex agnus* castus extract. The extracts are stressed under drastic conditions and the changes in the flavonoid content are investigated

(Figure 2.6). Flavonoids are separated on HPTLC silica gel 60F254 plates with tetrahydrofuran-toluene formic acid-water 16 : 8 : 2 : 1, v/v and detected by dipping of the warm plate into natural products reagent, followed by polyethylene glycol 400 solution. The results show that the flavonoids are rather stable under stress conditions. The concentrated HCl and hydrogen peroxide changes all other constituents. Figure 2.7 represents the quantified HPTL spectrum of a standard and mixture of quercetin, rutin and coumaric acid (present in the standard). Based on the peak area, the amount of each substance can be quantified.

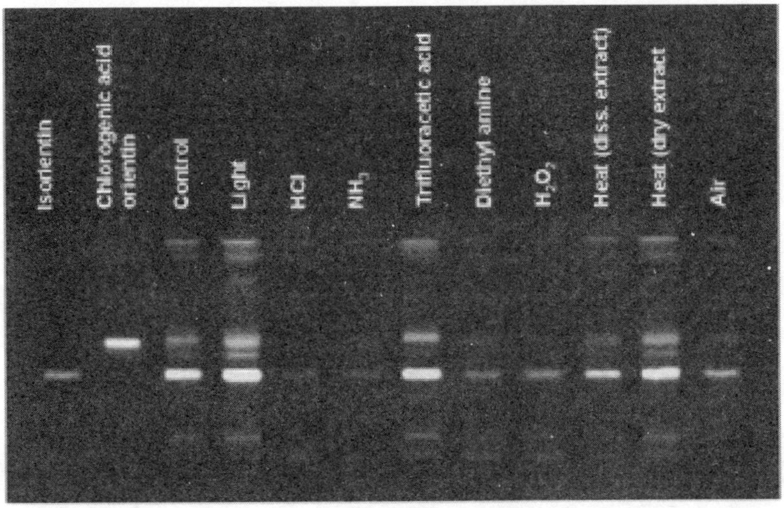

Figure 2.6 Results of stress tests on flavonoids (UV 366 nm)[27]

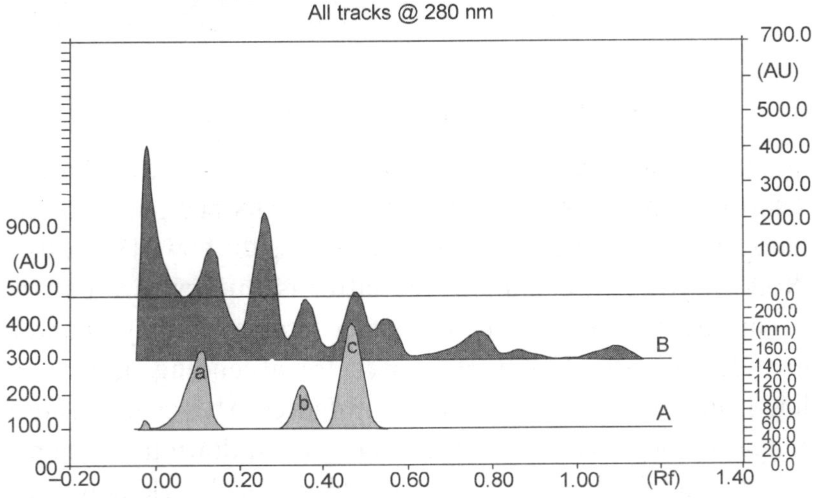

Figure 2.7 A HPTLC 3D overlay chromatogram of standard track (A) and sample track (B). Peaks a, b, and c represent quercetin, rutin and coumaric acid, respectively

PAPER CHROMATOGRAPHY

Particularly valuable separation results can be achieved when using various mobile phase systems to benefit from different separation mechanisms. For example, with cellulose one can apply a nonaqueous mobile phase to achieve the adsorption mechanism of retention and an aqueous mobile phase to achieve the partition mechanism.

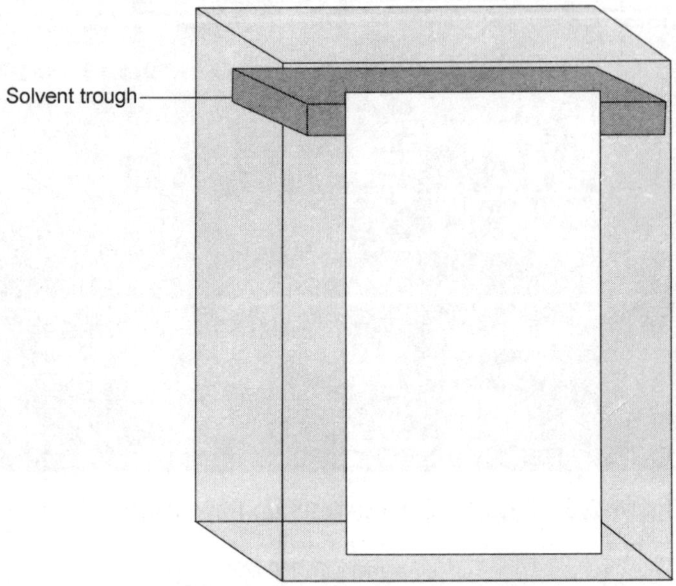

Figure 2.8 A sketch of paper chromatography set-up

Paper chromatography is usually carried out in a large glass tank (Figure 2.8) or cabinet and involves either ascending or descending flow of the mobile-phase solvents. Descending paper chromatography is faster due to gravity facilitating the flow of solvents. Large sheets of Whatman No.1 or No.2 filter paper (the latter is thicker) are cut into long strips (e.g., 22 × 56 cm long) for use in descending paper chromatography, or a wide strip of paper (e.g., 25 cm wide) of variable height is used for ascending paper chromatography. For descending-liquid paper chromatography, substances to be separated are applied as spots (e.g., 25 mm apart) along a horizontal pencil line placed down from the paper folded into V-shape. The V-trough folded paper is placed in a glass trough, held down by a glass rod, and when the tank has been equilibrated (vapour-saturated) with running solvents (mobile phase), the same solvent is added to the trough via a hole in the lid covering the chromatography tank.

The lid is sealed onto the chamber with stopcock grease in order to make the chamber airtight. After the mobile phase trails to the base of and off the paper sheet, the paper is hung to dry in a fume hood, where it can then be sprayed with reagents (e.g., ninhydrin reagent for amino acids) that give colour to the separated compounds of interest in white or UV light. Some compounds of interest have their own distinctive colours (e.g., chlorophylls), and hence, can be purified using this technique. In other cases, the dyes used to stain the location of the compound or protein cause irreversible covalent changes to the compound. In these cases, purification is not possible. In ascending paper chromatography, the same basic set-up and principles apply, with the exception that the mobile phase is placed at the bottom of the tank. Separation is achieved when the mobile phase travels up the paper via capillary action.

Solvents for Paper Chromatography

Solvents for paper chromatography include formamide, methanol, *n*-butanol, acetic acid, *n*-amyl alcohol, ethanol, isopropanol, acetone, *n*-propanol, etc.

In the case of paper chromatography, the stationary phase is not cellulose but the water that is held onto the paper, and hence is a partition between the water and the mobile phase. The paper takes up moisture when it is suspended in the closed chamber where the atmosphere is saturated with water vapour. It is a partition chromatography. Other polar stationary phases can be alcohol, formamide, propylene glycol, and so forth. This is normal chromatography.

In reverse phase paper chromatography, paraffin oil or silicone oil is used as the stationary phase. The paper is impregnated with the oil by using petroleum ether as the diluent. Alternatively, the hydrophilic nature of the paper can be modified chemically by esterification. The paper will then accept hydrophobic solvents from the atmosphere of the chamber.

Paper chromatography can be carried out in four different modes namely

- Ascending mode, where the mobile phase flows upward.
- Descending mode, where the mobile phase flows downward.
- Horizontal mode, where the mobile phase flows horizontally.
- Radial mode, where the mobile phase flows radially.

In a similar way, with the polar chemically bonded stationary phases, one can use nonaqueous mobile phases to achieve the adsorption mechanism of retention and the aqueous mobile phases to achieve the reversed-phase mechanism. Shifting from the adsorption to the partition mode causes marked differences in the separation selectivity. Paper chromatography is useful for very polar compounds.

Reagent	Compounds that can be identified	Observation
Iodine vapours	Organic molecules that are unsaturated	Brown spots
2´,7´-dihydrofluorescine	Organic molecules	Greenish yellow spots under UV light (254 nm)
Bromocresol green	Carboxylic acids	Yellow spots
50% Antimony trichloride in acetic acid	Steroids, alicyclic vitamins, carotenoids	Various colours
Ninhydrin	Amino acids	Purple colour

COLUMN CHROMATOGRAPHY

In column chromatography, the stationary phase, a solid adsorbent, is placed in a vertical glass (usually) column and the mobile phase, a liquid, is added to the top and flows down through the column (by either gravity or external pressure) (Figure 2.9). Column chromatography is generally used as a purification technique: it isolates desired compounds from a mixture.

Column chromatography consists of a column of particulate material such as silica or alumina that has a solvent passed through it at atmospheric or low pressure. The separation can be liquid/solid (adsorption) or liquid/liquid(partition).

There are four major descriptors that play a deciding role in the success of separation using any column namely retention factor, efficiency, selectivity and resolution. In column chromatography, a glass column of appropriate dimension, which is decided based upon the amount of the extract to be loaded is used. Poor separation would result, in case of bad selection of the column size. The column is packed with stationary phase (e.g., silica gel), either by dry packing or wet packing. The nature of the adsorbent and the solvent system should be decided based on the TLC profile. In case there are two compounds to be separated varying much in their R_f factor, dry packing would be acceptable for separation. The concept involved in this, is that tighter the packing, more the number of theoretical plates and better the separation. The resolution quality is very much proportional to the amount of extract loaded. Based on the amount of extract, the appropriate glass column can be separated (Figure 2.10).

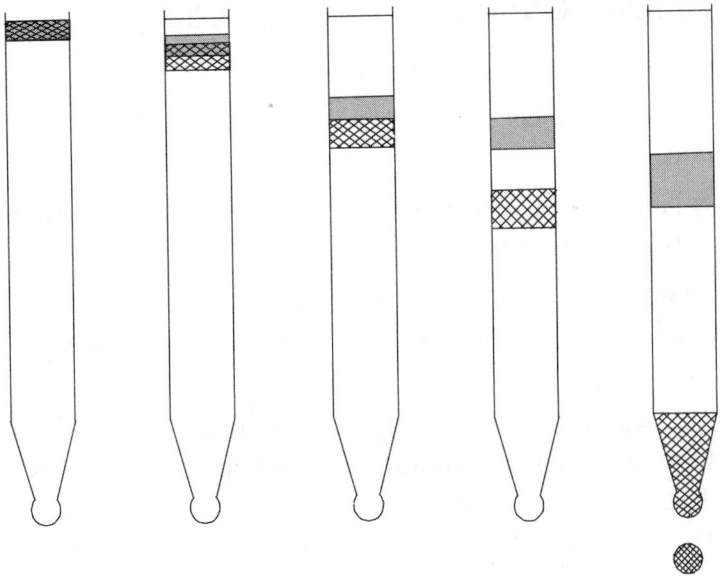

Figure 2.9 The process of elution in column chromatography

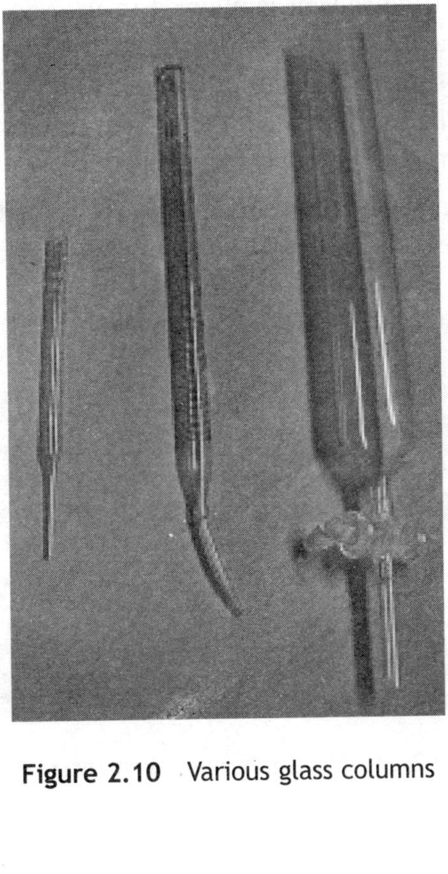

Figure 2.10 Various glass columns

How to Run a Column Chromatographic Separation

1. Make a slurry of silica gel in the solvent to be packed.
2. Pour it into the column taking care that air bubbles are not formed which might hinder good separation.
3. The extract is admixed with silica gel and loaded as a dry powder onto the top of the column and capped with cotton.
4. The column is eluted with the solvent and the fractions collected below and simultaneously monitored with TLC for the components present in each test tube.
5. If the compounds separated in a column chromatography procedure are coloured, the progress of the separation can simply be monitored visually. More commonly, the compounds to be isolated from column chromatography are colourless. In this case, small fractions of the eluent are collected sequentially in labelled tubes and the composition of each fraction is analysed by thin-layer chromatography.

Variations on Column Chromatography

Column chromatography can be categorized into two categories, depending on how the solvent flows down the column. If the solvent is allowed to flow down the column by gravity, or percolation, it is called gravity column chromatography. If the solvent is forced down the column by positive air pressure (or flushed with nitrogen), it is called flash chromatography. If vacuum is applied down the outlet to suck the solvent to move down faster, the method is know as vaccum liquid chromatography (Figure 2.11).

Flash chromatography and preparative HPLC have been used to separate the glucosinolates in an extract of broccoli seeds.[28, 29] High-speed countercurrent chromatography (HSCCC) is

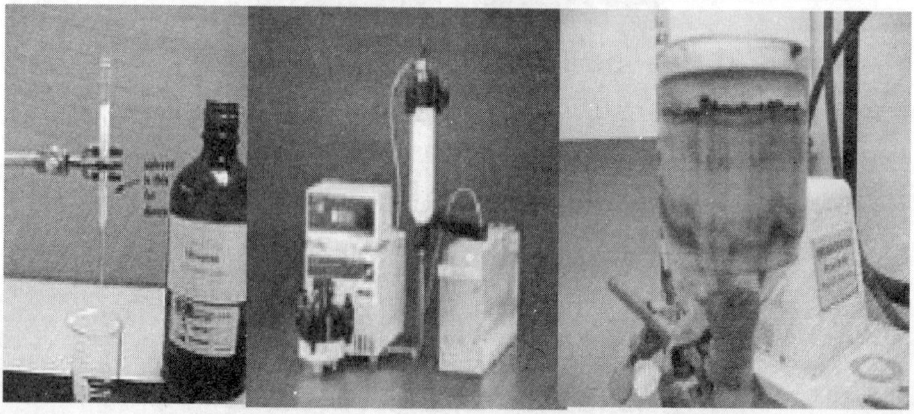

Figure 2.11 Variations in column chromatography

an all-liquid chromatographic system, working without solid support, in which separation phases occur.[30] Irreversible adsorption and artifact formation are thus minimized. Application of HSCCC in natural product chemistry is steadily increasing because of this advantage and excellent recoveries have been achieved.[31]

GEL PERMEATION CHROMATOGRAPHY—MOLECULAR SIEVES CHROMATOGRAPHY

Size-exclusion chromatography/gel filtration/permeation chromatography[32, 33] involves the use of porous gel molecules of agarose, cross-linked dextran, or polymers of acrylamide that allow the separation of compounds based on their molecular sizes. The pore size of the beads determines the molecular size range that can be separated with a particular column. Size separation is achieved by differential pore permeation. Below a certain size, a molecule can freely enter the holes in the stationary phase.

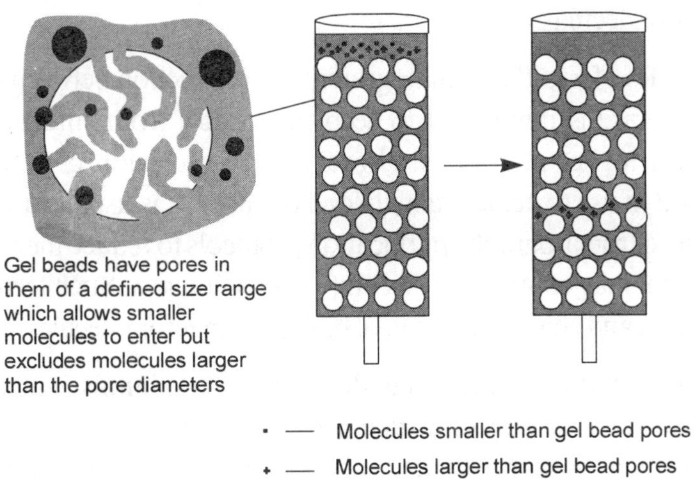

Figure 2.12 The process of gel permeation chromatography

Above a certain size, the molecule cannot enter the pores at all and is completely excluded. Molecules between these two size limits will be able to penetrate the pores more or less deeply and thus will be separated based on their sizes. As a result, the larger molecules passing from the column are eluted earlier than the smaller ones (Figure 2.12).

The analytes should not interact with the stationary phase and are separated only by their ability to penetrate the pores of the matrix. Buffers need to be chosen based on their ability to dissolve the sample and keep the analytes separated. Buffers mainly contain dissociating agents such as urea, sodium dodecyl sulphate, ethylene diamine tetra acetic acid, mercaptoethanol, dithiothreitol, etc., which prevent aggregation. Choice of column packing material is crucial and the pore size decides which molecular range can be separated. The resolution power of GPC as compared to most of the techniques is generally poor and the technique is mainly used for purification procedures and sample preparation.

The general rationale for separation is as follows:

1. A gel with an appropriate pore size is chosen in relation to the size of the molecule of interest.
2. Samples are added to the top of the gel column and are washed through, using a mobile phase that completely dissolves the molecules of interest.
3. Molecules that are too big to fit in the pores of the solid support will travel straight through the gel, and hence, elute from the column first.
4. Molecules that fit in the pores will penetrate the pores, which results in a longer path and elution at later time points. As a consequence, molecules normally elute in order of decreasing size.

Unlike other modes of liquid chromatography, the interaction between sample molecules and the stationary phase is minimized to keep the sample from spreading out within the column, resulting in higher recoveries. However, size-exclusion chromatography is not considered a method to use to achieve a high level of purity. It is usually used only as a clean-up step during liquid chromatography purification protocols to reduce the amount of larger or smaller molecules that may contaminate a compound of interest. Such techniques are widely used for the separation and characterization of biological macromolecules, especially proteins.

All size-exclusion separations are run under isocratic conditions, with a constant mobile phase.[34] Three major factors affecting such separations are the ionic strength, detergent concentration, and organic modifiers that may be added to the mobile phase. These conditions will also control which type of gel resin can be used. Consequently, a number of commercial gel matrices are available, such as Sephadex, Sepharose, Sephacryl, Sepharose CL, bio-gel,etc., each with different chemical properties and size-exclusion ranges.

HIGH-PERFORMANCE LIQUID CHROMATOGRAPHY

High-performance liquid chromatography is a popular method for the analysis of phytochemicals because it is easy to learn and use and is not limited by the volatility or stability of the sample compound. High-performance liquid chromatography (HPLC) is a form of liquid chromatography to separate compounds that are dissolved in solution.

Concept of Separation in HPLC

Compounds are separated by injecting a plug of the sample mixture onto the column. The different components in the mixture pass through the column at different rates due to differences in their partitioning behaviour between the mobile liquid phase and the stationary phase. The commonly used solvents include acetonitrile, methanol, water, etc. It is important that the solvents are degassed (removal of dissolved gases) which otherwise would damage the column. The column can be packed with silical gel (normal phase) or a nonpolar substance like octyl (C_8) or octadecyl silane (C_{18}) (reverse phase column). The difference between the normal phase and reverse phase column is that while in the former, due to the nature of the stationary phase non-polar compounds elute first, in the latter, polar compounds elute first.

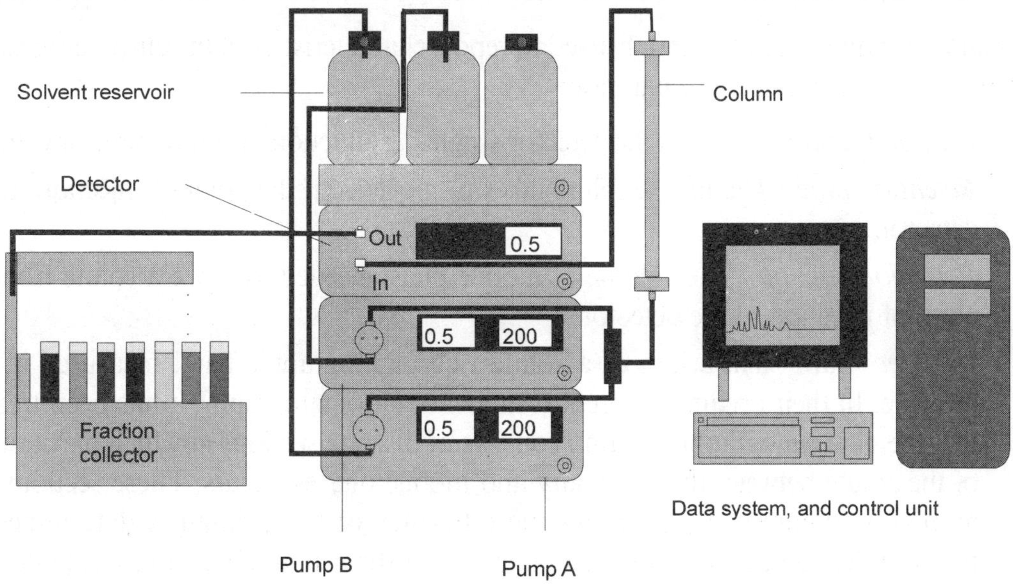

Figure 2.13 A schematic diagram of HPLC instrument

In cases where enantiomeric mixtures need to be separated, a chiral column is used which contains a chiral stationary phase which selectively separates the mixture to R and S.

HPLC instruments consist of a reservoir of mobile phase, a pump, an injector, a separation column, and a detector (Figure 2.13).

The chromatographic method most commonly used for analytical purposes is reversed-phase. Preparative HPLC refers to the process of isolation and purification of compounds. Important is the degree of solute purity and the throughput, which is the amount of compound produced per unit time. This differs from analytical HPLC, where the focus is to obtain information about the sample compound. The information that can be obtained includes identification, quantification, and resolution of a compound.

Each compound has a characteristic peak (Retention time—R_t) under certain chromatographic conditions. Depending on what needs to be separated and how closely related the samples are, the chromatographer may choose the conditions, such as the proper mobile phase, to allow adequate separation in order to collect or extract the desired compound as it elutes from the stationary phase. The migration of the compounds and contaminants through the column needs to differ enough so that the pure desired compound can be collected or extracted without incurring any other undesired compound.

Resolution of a Mixture of Two Compounds to Separate Entities and Elution

Four major descriptors are commonly used to report characteristics of the chromatographic column, system, and particular separation:

1. *Chromatogram* A plot of the detector's signal as function of elution time or volume.
2. *Retention time* The time a solute takes to move from the point of injection to the detector.
3. *Retention volume* The volume of mobile phase needed to move a solute from its point of injection to the detector.
4. *Baseline width* The width of a solute's chromatographic band measured at the baseline. In their original theoretical model of chromatography, it has been treated that the chromatographic column consists of discrete sections at which partitioning of the solute between the stationary and mobile phases occurs. These sections are termed as **theoretical plates** and the efficiency of the column is determined in terms of the number of theoretical plates, *N*, or the height of a theoretical plate.

Classification of HPLC Chromatographic Separations

HPLC techniques can be categorized into normal-phased, reverse-phased, ion exchange and size exclusion separations. The principal characteristic defining the identity of each technique is the dominant type of molecular interactions employed. There are three basic types of molecular forces: ionic forces, polar forces, and dispersive forces. Each specific technique capitalizes on each of these specific forces:

Normal phased HPLC Polar forces are the dominant type of molecular interactions employed in normal-phase HPLC. Normal-phase HPLC explores the differences in the strength of the polar interactions of the analytes in the mixture with the stationary phase. The stronger the analyte–stationary phase interaction, the longer the analyte retention. As with any liquid chromatography technique, NP-HPLC separation is a competitive process. Analyte molecules compete with the mobile-phase molecules for the adsorption sites on the surface of the stationary phase. The stronger the mobile-phase interactions with the stationary phase, the lower the difference between the stationary-phase interactions and the analyte interactions, and thus lower the analyte retention.

Mobile phases in NP-HPLC are based on nonpolar solvents (such as hexane, heptane, etc.) with the small addition of polar modifier (i.e., methanol, ethanol). Variation of the polar modifier concentration in the mobile phase allows for the control of the analyte retention in the column. Typical polar additives are alcohols (methanol, ethanol, or isopropanol) added to the mobile phase in relatively small amounts. Since polar forces are the dominant type of interactions employed and these forces are relatively strong, even only 1% v/v variation of the polar modifier in the mobile phase usually results in a significant shift in the analyte retention. Packing materials traditionally used in normal-phase HPLC are usually porous oxides such as silica (SiO_2) or alumina (Al_2O_3). The surface of these stationary phases is covered with a dense population of OH groups, which makes these surfaces highly polar. Analyte retention on these surfaces is very sensitive to the variations of the mobile-phase composition. Dispersive forces play a deciding role in reversed-phase HPLC . As opposed to normal-phase HPLC, reversed-phase chromatography employs mainly dispersive forces (hydrophobic or van der Waals interactions). The polarities of mobile and stationary phases are reversed, such that the surface of the stationary phase in RP-HPLC is hydrophobic and mobile phase is polar, where mainly water-based solutions are employed. Reversed-phase HPLC is by far the most popular mode of chromatography. Almost 90% of all analyses of low-molecular-weight samples are carried out using RP-HPLC. One of the main drivers for its enormous popularity is the ability to discriminate very closely related compounds and the ease of variation of retention and selectivity.

Ion-exchange HPLC is governed by ionic forces.

In case of IE-HPLC, four major types of ion-exchange centres are usually employed:

1. SO_3^-—strong cation-exchanger
2. CO_2^-—weak cation-exchanger
3. Quaternary amine—strong anion-exchanger
4. Tertiary amine—weak anion-exchanger

The fourth type of HPLC technique, size-exclusion HPLC, is based on the absence of any specific analyte interactions with the stationary phase (no force employed in this technique). SEC is the method for dynamic separation of molecules according to their size; as indicated by its name, the separation is based on the exclusion of the molecules from the porous space of the packing material due to their steric hindrance.

AFFINITY CHROMATOGRAPHY

Affinity chromatography is defined as a liquid chromatographic technique that makes use of a "biological interaction" for the separation and analysis of specific analytes within a sample. Examples of these interactions include the binding of an enzyme with an inhibitor or of an antibody with an antigen. Such binding processes are used in affinity chromatography by first obtaining a binding agent known as the "affinity ligand", that selectively interacts with the desired analyte and then placing this ligand onto a solid support within a column. Once this immobilized ligand has been prepared, it can be used for isolation or quantification of the analyte.

The immobilized ligand is the key factor that determines the success of any affinity chromatographic method. As implied by the definition given earlier for affinity chromatography, most of these ligands are of biological origin; however, the term "affinity chromatography" has also been used throughout the years to describe some columns that contain selective ligands of nonbiological origin. Examples of these nonbiological ligands are boronates, immobilized metal ion complexes, and synthetic dyes (e.g., triazine-related compounds). Terms such as "bioaffinity chromatography" and "biospecific adsorption" are occasionally used to specify whether the affinity ligand is really a biological compound. Regardless of the origin of the ligand, the type of ligand can be used to divide affinity techniques into various subcategories, such as lectin, immunoaffinity, dye ligand, and immobilized metal ion affinity chromatography, to name a few [35, 36]

COUNTERCURRENT CHROMATOGRAPHY (CCC)

CCC is very effective for initial fractionation of crude plant extracts. It can be used for all ranges of polarities, but has special advantages for the handling of polar extracts, which are often difficult to process by more classical techniques. A big problem when dealing with crude extracts of plants is the fractionation of large amounts of sample without suffering too much material loss. Solid supports, and most of all, silica gel, are notorious for irreversibly absorbing large proportions of sample. When the absorbed material is the required bioactive component of a sample, the complications are evident. Countercurrent chromatography provides a means of avoiding this problem.

Countercurrent Chromatography (CCC) was developed by Ito in 1960.[37, 38] Droplet countercurrent chromatography (DCCC), centrifugal partition chromatography, (CPC), high-speed countercurrent chromatography (HSCCC), and elution extrusion countercurrent chromatography (EECCC)—all these techniques are based on the partitioning of a component in two nonmiscible liquids and can be classified as liquid chromatographic techniques. The unique feature of CCC is that, unlike all other techniques, the stationary phase is also a liquid without any solid support and is based on a coil planet centrifuge system. Thus, this technique is an all-liquid method without solid phases, which relies on the partition of a sample between two immiscible solvents to achieve separation.

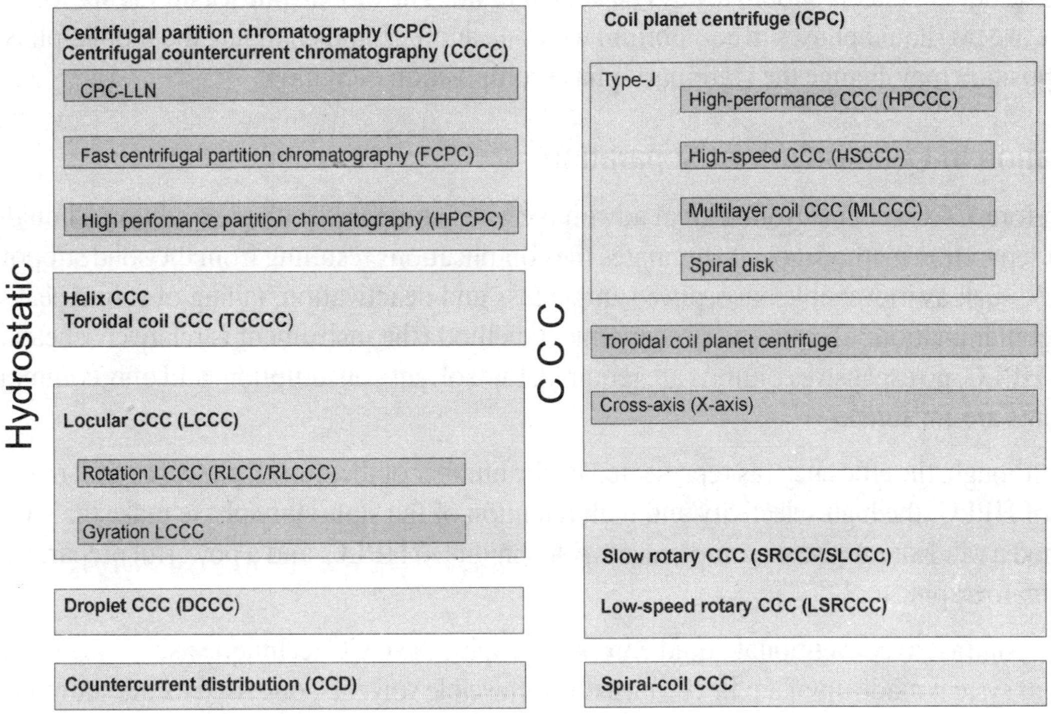

Figure 2.14 The varying principles in countercurrent separation

A typical CCC set-up comprises a mobile phase reservoir, a pump, an injection valve, a column, a detector, and a recorder or data processor. This is the exact description of any HPLC unit. The salient feature in this type of chromatography is that main "columns" contain rotating parts needed to create and maintain a centrifugal field that will work with a support-free liquid stationary phase.

The relative proportion of solute passing into each of the two phases is determined by the respective partition coefficients.

Liquid–liquid partition is the only phenomenon responsible for solute retention in CCC. However, CCC is not an automated liquid–liquid extraction process in the same way that HPLC is not an automated solid phase extractor. In the chromatographic process, the solutes undergo a number of back-and-forth liquid-phase exchanges. The plate count is a theoretical measure of this number. The solutes separate because of their differing affinity for the stationary phase (Figure 2.14).

A biphasic liquid system is used and the more polar liquid phase or the less polar liquid phase can be used as the stationary phase. The phases' roles can even be interchanged during the chromatographic separation—this is called dual-mode CCC. The physical chemistry of biphasic liquid systems should always be kept in mind. The mobile phase and the stationary phase are two liquid phases in equilibrium with one another, and changing the mobile-phase composition may change the stationary-phase composition even more.

Variation in Countercurrent Separation

Therefore, CCC benefits from a great advantage when compared with the traditional liquid–solid separation methods: (1) it eliminates the complications resulting from the solid support matrix, such as irreversible adsorptive sample loss and deactivation, tailing of solute peaks, and contamination; (2) it is a very economical method (the instrument is relatively cheaper than HPLC, no expensive columns are required, low solvent consumption, and only common solvents are consumed).

Although, the efficiency (as represented by the number of theoretical plates) cannot match that of HPLC, the high selectivity and high retention of the stationary phase make the CCC method a valid alternative or complementary technique to HPLC, and a powerful preparative chromatographic tool. [7]

In contrast to conventional liquid chromatography, the CCC technique uses a two-phase solvent system made up of a pair of mutually immiscible solvents, one used as the stationary phase and the other as the mobile phase. The use of two-phase solvent systems allows one to choose solvents from an enormous number of possible combinations. Therefore, the selection of a suitable two-phase solvent system is the key element in the CCC method development. However, the main difficulty also arises from the choice of the solvent in an enormous number of possibilities available to the analyst.

The CCC method provides an advantage over the conventional column chromatography by eliminating the use of a solid support where an amount of stationary phase is limited, and dangers of irreversible adsorption from the support are inevitably present. Thus, both crude plant extracts and semipure fractions can be chromatographed, with sample loads ranging from milligrams to grams. Furthermore, the use of two-phase solvent systems allows one to choose solvents from an enormous number of possible combinations, which enables the CCC technique to separate compounds with a wide range of polarities. As a result, CCC comes to be a powerful chromatographic tool for preparative isolation and purification of bioactive compounds from natural sources. Moreover, the production of active compounds or fractions can be used as pure reference standards for further biological, pharmacological,[39] and clinical studies.

1. CCC can cope with a wide range of radically different polarity compounds.
2. CCC can cope with particulates and extract solid samples.
3. CCC can have orthogonal selectivity to HPLC.
4. CCC can achieve 99+% purity of target compounds from complex samples.
5. CCC can take extremely complex matrices, such as natural product extracts and heart cut target polarities, bio-actives, etc. without risk of on-column degradation or adsorption.
6. CCC uses exactly the same type of liquid pumps, injectors, switching valves fraction collectors, etc. as HPLC, flash chromatography or SMB, but as CCC often requires only a fraction of the solvent flow or total solvent used, it offers substantial cost savings in capital equipment cost by using lower flow pumps, plus savings in solvent and process cost.
7. Scale-up is linear and predictable for the modular Quattro CCC.
8. Small laboratory 100 ml coil CCC can rapidly process 0.5 to 2+ grams of crude extract, but requires flows of only 2 to 10 ml/min and therefore can use a standard analytical HPLC pump.
9. Single process units of 3000 ml can rapidly prepare 15 to 60+ grams of crude extract, but requires flows of only 25 to 100 ml/min. HPLC pumps of this capacity are available at modest cost.
10. Any number of process units can be used in series/manifolded in parallel/or used in SMB mode for tonne per annum production.

Gas Chromatography

It is well-known that many pharmacologically active components in natural products are volatile chemical compounds. The advantages of GC clearly lie in its high sensitivity of detection for almost all the volatile chemical compounds. This is especially true for the usual FID detection and GC–MS. Furthermore, the high selectivity of capillary columns enables separation of many volatile compounds simultaneously within comparatively short times. Thus, over the past decades, GC is a popular and useful analytical tool in the research field of natural products.[40–49]

Gas chromatography is the use of a carrier gas to convey the sample (as a vapour) through a column consisting of an inert support and a stationary phase that interacts with sample components. Thus it is usually partition chromatography. There are also a range of materials, especially for permanent gas and light hydrocarbon analysis that utilize adsorption. The simplest partition systems consisted of a steel tube filled with crushed brick that had been coated with a hydrocarbon that had a high boiling point, e.g., squalane. Today, the technique uses very narrow fused silica tubes (0.1 to 0.3 mm ID) that have sophisticated stationary phase films (0.1–5 μm) bonded to the surface and also cross-linked to increase thermal stability.

The ability of the film to retard specific compounds is used to ascertain the "polarity" of the column. If benzene elutes between normal alkanes where it is expected by boiling point (midway between *n*-hexane and *n*-heptane), then the column is "non-polar," e.g., squalane and methyl silicones. If the benzene is retarded until it elutes after *n*-dodecane, then the column is "polar" e.g., OV-275 (dicyanoallyl silicone) and 1, 2, 3-tris (2-cyanoethoxy) propane. In general, polar columns are less tolerant of oxygen and reactive sample components, but the ability to select different polarity columns to obtain satisfactory peak resolution is what has made GC so popular.

The column is placed in an oven that has exceptional temperature control, and the column can be slowly heated up to 350–450°C (sometimes starting at –50°C to enhance resolution of volatile compounds) to provide separation of wide-boiling range compounds. The carrier gas is usually hydrogen or helium, and the eluting compounds can be detected several ways, including flames (flame ionization detector), by changes in properties of the carrier (thermal conductivity detector), or by mass spectrometry. The availability of "universal" detectors such as the FID and MS, makes GC a popular tool in laboratories handling organic compounds. There are also columns that have a layer of 5–10 μm porous particulate material (such as molecular sieve or alumina) bonded to the inner walls (PLOT—Porous layer open tubular), and these are used for the separation of permanent gases and light hydrocarbons. GC is restricted to molecules (or derivatives) that are sufficiently stable and volatile to pass through the GC intact at the temperatures required for the separation. Specialist books on the production of derivatives for GC are available.

The mixture to be separated and analysed may be either a gas, a liquid, or a solid in some instances. All that is required is that the sample components be stable, have a vapour pressure of approximately 0.1 Torr at the operating temperature, and interact with the column material (either a solid adsorbent or a liquid stationary phase) and the mobile phase (carrier gas). The result of this interaction is the differing distribution of the sample components between the two phases, resulting in the separation of the sample component into zones or bands. The principle that governs the chromatographic separation is the foundation of most physical methods of separation, for example, distillation and liquid–liquid extraction. Separation of the sample components may be achieved by one of three techniques: frontal analysis, displacement development, or elution development.

Frontal Analysis

The liquid or gas mixture is fed into a column containing a solid packing. The mixture acts as its own mobile phase or carrier, and the separation depends on the ability of each component in the mixture to become a sorbate. Once the column packing has been saturated (i.e., when it is no longer able to sorb more components), the mixture then flows through with its original composition.

The early use of this technique involved measurement of the change in concentration of the front leaving the column; hence the name "frontal analysis."

The least-sorbed component breaks through first and is the only component to be obtained in a pure form.

Displacement Development

In this technique the developer is contained in the moving phase, which may be a liquid or a gas. One necessary requirement is that this moving phase be more sorbed than any sample components. One always obtains a single pure band of the first component in the sample. In addition, there is always an overlap zone for each succeeding component, which is an advantage of this technique over frontal analysis. The disadvantage, from the analytical viewpoint, is that the component bands are not separated by a region of pure mobile phase.

Elution Development

In this technique, components A and B travel through the column at rates determined by their retention on the solid packing sorption are sufficient or the column is long enough, a complete separation of A and B is possible. Continued addition of eluent causes the emergence of separated bands or zones from the column. A disadvantage of this technique is the very long time interval required to remove a highly sorbed component. This can be overcome by increasing the column temperature during the separation process.

REFERENCES

1. Tswett, M. (1906). "Works of the Warsaw society of natural science." *Ber. Deut. Bot. Ges. XXIV* 316.
2. Tiselius, A. (1940). *Ark. Kemi. Mineral. Geol.* 14B(22) NA.

3. Martin, A.J.P. and Synge, R.L.M. (1941). "A new form of chromatogram employing two liquid phases." *Biochem. J. (Lond.)* 35, 1358.
4. Consden, R., Gordon, A. H. and Martin, A. J. P. (1944). "Qualitative analysis of protein: a partition chromatographic method using paper." *Biochem. J.* 38(3), 224–232.
5. Cremer, E. and Prior, F. (1951). Z. *Elektrochem.* 55, 66 (1951). Cremer, E. and Muller, R. (1951). *Z. Elektrochem.* 55, 217.
6. James, A.T. and Martin, A.J.P. (1952). "Gas-liquid partition chromatography: the separation and micro-estimation of volatile fatty acids from formic acid to dodecanoic acid." *Biochem. J.* 50, 679–690.
7. Giddings, J.C. (1965), "*Dynamics of Chromatography, Part I,*" *Principles and Theory.* Marcel Dekker, New York. pp.13–26.
8. Pecosk, R.S., Shields, L.D., Cairns, T. and McWilliam, I.G. (1976). *Modern Methods of Chemical Analysis.* Wiley, New York.
9. Wagner, H. and Bladt, S. (1996). *Plant Drug Analysis: A Thin Layer Chromatography Atlas,* 2nd edn. Springer-Verlag, Berlin Heidelberg, New York, Tokyo. p. 384.
10. Skoog, D., West, D. and Holler, J. (1988). *Fundamentals of Analytical Chemistry,* 5th edn. Saunders College of Publishing. A division of Holt, Rinehart and Winston. New York-Chicago-San Fransisco. p. 894.
11. Fried, B. and Sherma, J. (1994). *Thin-layer Chromatography: Techniques and Applications,* 3rd edn., *Chromatographic Science Series* 66. Marcel Dekker, Inc. New York, USA. p. 451.
12. Fritz, J. and Schenk, G. (1987). *Quantitative Analytical Chemistry.* Allyn and Bacon, Inc. Boston-London-Sydney-Toronto. p. 690.
13. Abidi, S.L. (1998). "Separation procedures for phosphatidylserines." *J. Chromatogr. B.* 717(1–2), 279–293.
14. Lawton, L.A. and Edwards, C. (2001). "Purification of microcystins." *J. Chromatog.* 912(2), 191–209.
15. Bhushan, R. and Martens, J. (2001). "Separation of amino acids, their derivatives and enantiomers by impregnated TLC." *Biomed. Chromatogr.* 15(3), 155–165.
16. Reddy, M.V. (2000). "Methods for testing compounds for DNA adduct formation." *Reg. Toxicol. Pharmacol.* 32(3), 256–263.
17. Alemany, G. Akaarir, M. Gamundi, A. and Nicolau, M.C. (1999). "Thin-layer chromatographic determinations of catecholamines, 5-hydroxytryptamine, and their metabolites in biological samples—a review." *J. AOAC Int.* 82(1), 17–24.

18. Roda, A. Piazza, F. and Baraldini, M. (1998)." Separation techniques for bile salts analysis." *J. Chromatogr. B.* 717(1–2), 263–278.

19. Porter, J.K. (1995). "Analysis of endophyte toxins: fescue and other grasses toxic to livestock." *J. Animal Sci.* 73(3), 871–880.

20. Oka, H., Ito, Y., and Matsumoto, H., (2000). "Chromatographic analysis of tetracycline antibiotics in foods." *J. Chromatogr.* 882(1–2), 109–133.

21. Sherma, J. (2000). "Thin-layer chromatography in food and agricultural analysis." *J. Chromatogr.* 880(1–2), 129–147.

22. Muthing, J. (1996). "High resolution thin-layer chromatography of gangliosides." *J. Chromatogr. A.* 720(1–2), 3–25.

23. Bereznitski, Y., Thompson, R.O' Neill, E. and Grinberg, N. (2001). "Thin-layer chromatography—a useful technique for the separation of enantiomers." *J. AOAC Int.* 84(4), 1242–1251.

24. Mengning lan, dongyuan wang, wei wei, wei lu. (2003). "Multidimensional relay development in TLC." *Journal of Planar Chromatography.* 16(6), 461–464.

25. Joseph Sherma and Bernard Fried (2003). *Handbook of Thin-layer Chromatography.* Marcel Dekker, New York.

26. Joseph, C., Touchstone. (1992). *Practice of Thin Layer Chromatography.* Wiley-Interscience, New York.

27. Wahli, F. Stability tests of *Vitex agnus castus* (Chaste tree) extracts. www.camag.com.

28. Arguello, L.G., Sensharma, D.K., Qiu, F., Nurtaeva, A. and Rassi, Z.E. (1999). "High-performance liquid phase separation of glycosides analytical and micropreparative HPLC combined with spectroscopic and enzymatic methods for generating a glucosinolate library" *J. AOAC Int.,* 82, 1115.

29. Rochfort, S., Caridi, D., Tinton, S.M., Trenerry, V.C. and Jones, R. (2006). *J. Chromatogr. A.* 1120, 205.

30. Ito, Y.and Conway W.D. (Eds.). (1996). *High-Speed Countercurrent Chromatography.* Wiley–Interscience, New York.

31. Du Q. and Ito, Y. (2005). "Review on scale-up of coil column counter–current chromatographs." *Curr. Pharm. Anal.* 1, 309.

32. Bruno, T.J. (1991). *Chromatography and Electrophoretic Methods.* Prentice-Hall, Englewood Cliffs, New Jersey.

33. Heftmann, E. (1992). *Fundamentals and Techniques in Chromatography,* 5th edn., Part A, *Journal of Chromatography Library, 51A.* Elsevier, New York.

34. Cseke, L., Kaufman, P., Podila, G.K. and Tsai, C.J. (2004). *Handbook of Molecular and Cellular Methods in Biology and Medicine*, 2nd edn., CRC Press, Boca Raton, Florida.

35. Hage, D.S. (1998). "Affinity chromatography." In: Katz, E., Eksteen, R., Shoenmakers, P. and Miller, N. (Eds.). *Handbook of HPLC*. Marcel Dekker, New York. pp. 483–98.

36. Hermanson, G.T., Mallia, A.K. and Smith, P.K. (1992). *Immobilized Affinity Ligand Techniques*. Academic Press, New York. p. 454.

37. Ito, Y., Weinstein, M., Aoki, I., Harada, R., Kimura, E.and Nunogaki, K. (1966). "The coil planet centrifuge." *Nature*. 212, 985–987.

38. Ito, Y. and Bowman, R.L. (1970). "Countercurrent chromatography: Liquid-liquid partition chromatography without solid support." *J. Chromatogr. Sci.* 8, 315–323.

39. Lu, Y., Sun, C. and Pan, Y. (2006). "A comparative study of upright countercurrent chromatography and high-performance liquid chromatography for preparative isolation and purification of phenolic compounds from *Magnoliae officinalis*." *J. Sep. Sci.* 29, 351–357.

40. Brochmann-Hanssen, R.and Baerheim Svendsen, A. (1962). "Gas chromatography of alkaloidal salts and derivatives." *J. Pharm. Sci.* 51,1095.

41. Majlat, P. (1982). "Gas-chromatographic assay of atropine and phenobarbitone in pharmaceutical preparations containing valerian liquid extract. "*J. Chromatogr.* 241(2), 399–403.

42. Briggs, C.J.and Simons, K.J. (1983). " Gas-chromatographic assay of atropine in formulations containing atropine sulphate and cholinesterase reactivators." *J. Chromatogr.* 257(1), 132–136.

43. Majlat, P. (1984). " Gas-chromatographic determination of atropine, theophylline, phenobarbitone and amidopyridine in tablets."*Pharmazie*. 39(5), 325–326.

44. Soleas, G.J., Diamandis, E.P., Karumanchiri, A.and Goldberg, D.M. (1997). "A multiresidue derivatization gas chromatography assay for fifteen phenolic constituents with mass selective detection." *Anal. Chem.* 69(21), 4405–4409.

45. Molyneux, R.J., Mahoney, N., Bayman, P., Wong, R.Y., Meyer, K. and Irelan, N. (2002). "Eutypa die back in grape Vines: Differential production of acetylenic phenol metabolites by strains of Eutypa lata." *J. Agric. Food Chem.* 50(6), 1393–1399.

46. Angerosa, F., d'Alessandro, N., Corana, F.and Mellerio, G. (1996). "Characterization of phenolic and secoiridoid aglycons present in virgin oil by gas chromatography-chemical ionization mass spectrometry." *J. ChromatogrA.* 736, 195.

47. Bunzel, M., Ralph, J., Marita, J.M. Hatfield, R.D.and Steinhart, H. (2001). "Diferulates as structural components in soluble and insoluble cereal dietary fibre." *J.Sci. Food Agric.* 81, 653.

48. El-Shazly, A. Tei, A. Witte, L. El-Domiaty, M.and Wink, M. (1997). "Tropane alkaloids of *Hyoscyamus boveanus, H. desertorum, H. Mutius and H. albus* from Egypt." *Zeitschrift fir Naturforschung; Journal of Bioscience.* 526, 729–739.

49. Ylinen, M., Naaranlahti, T. Lapinjoki, S.Huhtikangas, A. Salonen, M.L. Simola, L.K.and Lounasmaa, M. (1986). "Tropane alkaloids from *Atropa belladona*: Part I capillary gas chromatographic analysis." *Planta Med.* 52, pp. 85–86.

3

STRUCTURAL ELUCIDATION OF NATURAL PRODUCTS

INTRODUCTION

Natural products chemistry can be divided into three different fields: methods of isolation, methods of structure elucidation and synthetic approaches. Though in the last two decades numerous new types of chemical reactions have been found, the methods of isolation and structure elucidation have changed more fundamentally than those of synthesis.

Two decades ago, the structural elucidation of natural compounds were mainly based on chemical reactions like those of degradation and derivatization. Only few physico-chemical parameters were available for characterization, viz., melting points, solubilities, values of elemental analysis, molecular weights, or specific rotations. Since then different physicochemical instruments like those of absorption spectroscopy (ABS), optical rotatory dispersion (ORD), circular dichroism (CD), mass spectrometry (MS), and nuclear magnetic resonance (NMR) were developed which allow recording of spectra routinely and thus the basis of correlation between spectral characteristics and structure for organic molecules was laid.

This chapter gives a brief introduction of biophysical techniques that are widely used in the structural investigation of natural products. The principle/phenomena behind such techniques have been widely dealt in various books. The aim of this chapter is to highlight the utility of each technique.

The chemical structures of natural product compounds though very elegant in their nature are tremendously diverse.[1–4] Such diversity presents a challenge in unravelling the mystery of the chemical structure of an unknown material presented. Moreover, phytochemicals possess multiple chiral centres, which is essential for their bioactivity. Spectroscopic techniques play a determined role in the structural elucidation of natural products, and involves the study of the interaction of electromagnetic radiation (energy) and matter.

Figure 3.1 depicts the span of the electromagnetic spectrum with its characteristic subdivisions.

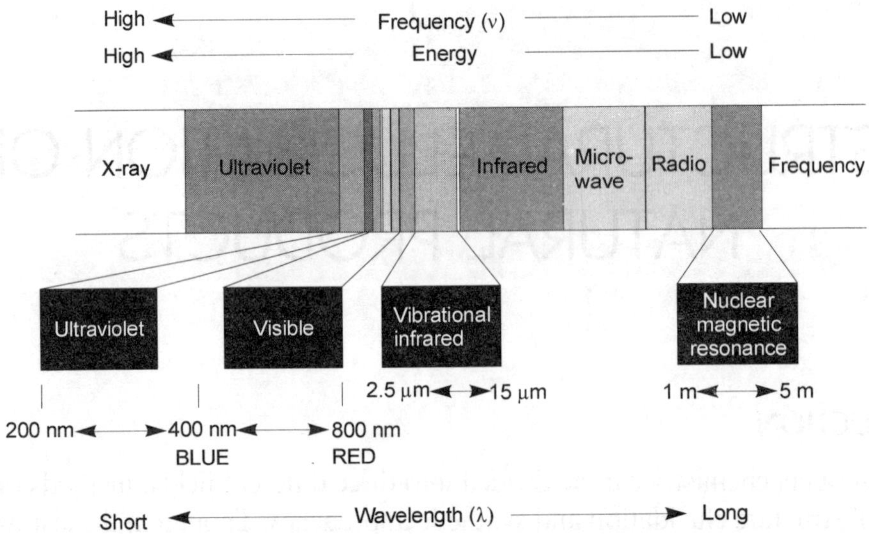

Figure 3.1 Electromagnetic spectrum and its characteristic wavelength

When energy is applied to matter it can be absorbed, be emitted, cause a chemical change (reaction) or be transmitted.

Atoms and molecules have a range of discrete energy levels corresponding to different electronic, vibrational or rotational states. The interaction between atoms and electromagnetic radiation is characterized by the absorption and emission of photons, such that the energy of the photons exactly matches an energy level difference in the atom. Since the energy of a photon is proportional to the frequency, the different forms of spectroscopy are often distinguished on the basis of the frequencies involved. For instance, absorption and emission between electronic states of the outer electrons typically require frequencies in the ultraviolet (UV) range, hence giving rise to UV spectroscopy. Molecular vibrational modes are characterized by frequencies just below visible red light and are thus studied with infrared (IR) spectroscopy. Nuclear magnetic resonance (NMR) spectroscopy uses radio frequencies, which are typically in the range of 10–800 MHz.

The concept in spectroscopic applications is as follows. All molecules are made of atoms which are intra-connected. They exhibit vibration motion: have electrons involved in their bonding and respond to incident radiation. Different electromagnetic vibrations elicit varying response from them. Gamma rays influence the nucleus, X-rays, the inner electrons, infrared, the molecular vibrations, microwaves, the molecular rotations, electron spin flips.

SPECTROSCOPY TECHNIQUES

Ultraviolet spectroscopy (UV) speaks about the electronic energy states and is used in the detection of conjugated molecule, carbonyl group, nitro group, etc.

Infrared spectroscopy (IR) investigates the vibrational energy states and is useful in the identification of functional groups.

Nuclear Magnetic Resonance (NMR) spectroscopy identifies the nuclear spin states and deciphers the number, type, and relative position of protons (hydrogen nuclei) and carbon-13 nuclei apart from some more nuclei.

Mass Spectrometry (MS) involves high-energy electron bombardment of the molecule of interest and is useful in the determination of molecular weight and of the presence of nitrogen, halogens, etc.

The usage of different techniques in structural elucidation can be likened to "blind men feeling the elephant" parlance. Each technique gives a unique information about the molecule, which when put together picturizes the complete structure.

The applications of the widely used biophysical techniques are given in Table 3.1.

Table 3.1 Biophysical techniques and the structural information derived from them

Technique	Structural information
IR, NMR, UV	Functional groups
MS^n, NMR	Building blocks or repeating units (e.g., amino acids, carbohydrates), structural formula
MS	Molecular weight, elemental composition
NMR	Relative stereochemistry
Circular dichroism, specific optical rotation	Absolute configuration
X-ray crystallography (compound must crystallize)	Structure with absolute configuration

Electron impact, chemical ionization, electrospray ionization, and laser desorption ionization are routinely used in conjunction with mass spectrometry to ionize components in mixtures. These ionization techniques result in different fragmentation patterns of molecules and when used collectively, the data can be utilized to identify molecular fragments, as well as overall molecular structure. Nuclear magnetic resonance spectroscopy is routinely used to determine the structural components of molecules. The technique is particularly useful for recognizing different configurations of molecules. NMR is also used to identify impurities (above 1%), quantitate trace components, determine the extent of reaction, evaluate purity, and measure the rates of kinetic processes.

THE STRUCTURE ELUCIDATION PROCESS

The following steps of the structural elucidation process should be considered to be targets amenable to automation:

1. The choice of the smallest group of procedures that is most likely to reveal the unknown structure (NMR, Fourier Transform-NMR, MS, MS/MS, IR, FTIR, ultraviolet (UV), etc.)
2. The acquisition of data from the selected procedures
3. The analysis of data from the selected procedures
4. The use of a computer program to construct the structure from collected spectroscopic and spectrometric data.

Historically, structure identification has occurred after purification was complete. However, not infrequently in these cases, structural identification revealed that all of the previous laborious steps of purification had produced a compound that was already known or of an undesirable type.

The identification of unwanted compounds or de-replication should occur as early as possible in the natural product isolation and purification process, to avoid the loss of time and funds in the case of sponsored projects. The development of coupled techniques has permitted the achievement of that goal. Indeed, coupled techniques such as LC/NMR/MS have now evolved and have potent application in the pharmaceutical field. As the sensitivity of instrumentation continues to improve, the value of these coupled techniques will increase even more.

ULTRAVIOLET SPECTROSCOPY

Ultraviolet/visible spectroscopy involves the absorption of ultraviolet/visible light by a molecule causing the promotion of an electron from a ground electronic state to an excited electronic state. The range is 190–800 nm. The types of electrons and their electronic transitions are given in Figure 3.2.

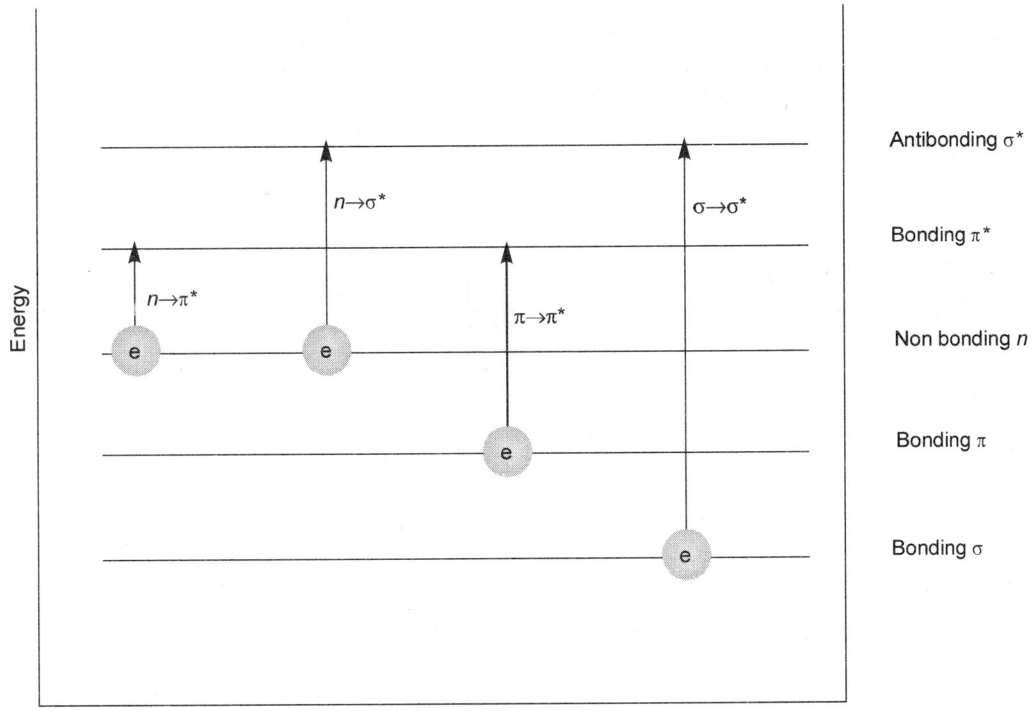

Figure 3.2 Electronic transition of interest in ultraviolet spectroscopy

Selective ultraviolet absorption arises from excitation of electronic energy levels associated with certain specific resonating groups called chromophores within molecules. The scope of ultraviolet spectroscopy is limited by the fact that numerous compounds do not exhibit selective absorption in the ultraviolet region of the spectrum. However, the limited scope of ultraviolet methods represents an advantage rather than a disadvantage when an ultraviolet-absorbing component is to be determined in the presence of non-ultraviolet-absorbing materials. Typical absorption of certain functional groups are given in Figure 3.3.

The ultraviolet spectrum, while to some extent characteristic of the individual molecule, serves primarily to establish compound class and is rather severely limited in its ability to distinguish a specific compound under study from all others. An infrared or Raman spectrum, on the other hand, uniquely characterizes an individual compound and serves also for the detection of many functional groups and atomic configurations which are not detectable by the ultraviolet method.

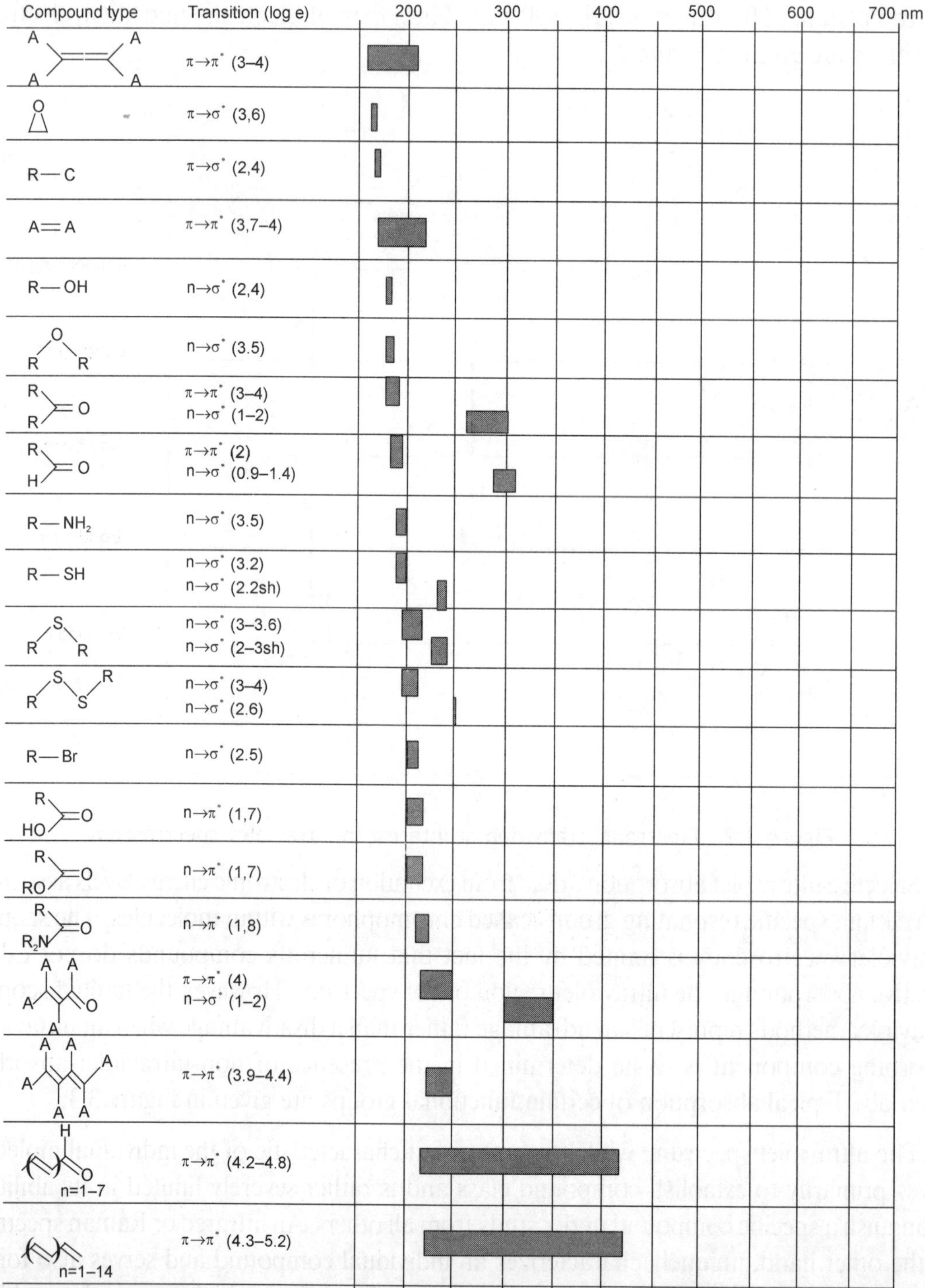

Figure 3.3 UV/V is absorption bands of various compound types (A: alkyl or H; R: alkyl;sh: shoulder)

INFRARED AND RAMAN SPECTROSCOPY OF NATURAL PRODUCTS[5, 6]

Both infrared and Raman spectra arise from the excitation of vibrational energy levels associated with the complex and unique set of interatomic vibration frequencies characteristic of a specific molecule.

The infrared region of the electromagnetic spectrum extends from 14,000 cm^{-1} to 10 cm^{-1}. The region of most interest for chemical analysis is the mid-infrared region (4,000 cm^{-1} to 400 cm^{-1}) which corresponds to changes in vibrational energies within molecules. The far infrared region (400 cm^{-1} to 10 cm^{-1}) is useful for molecules containing heavy atoms such as inorganic compounds but requires rather specialized experimental techniques. By contrast, all organic and many inorganic materials will exhibit analytically useful infrared or Raman spectra, if sampling considerations permit obtaining of such spectra.

An infrared spectrum measures the absorption of infrared radiation by vibrations of the atoms in a compound when those vibrations result in a change of dipole moment. The spectrum is a plot of the per cent of radiant energy absorbed by the sample as a function of the radiation frequency. In principle, each compound has a unique IR spectrum, which can be used like a fingerprint to identify it. In practice, instruments are not sensitive enough to distinguish between members of a homologous series which differ very little.

It is rarely, if ever, possible to identify an unknown compound by using IR spectroscopy alone. Its principal strengths are: (i) it is a quick and relatively cheap spectroscopic technique, (ii) it is useful for identifying certain functional groups in molecules and (iii) an IR spectrum of a given compound is unique and can therefore serve as a fingerprint for this compound.

The characteristic absorptions of certain functional groups are given in Table 3.2.

All molecules vibrate, even at a temperature of absolute zero. In general, a polyatomic molecule with N atoms has 3N–6 distinct vibrations (Figure 3.4). Each of these vibrations has an associated set of quantum states and in IR spectroscopy, the IR radiation induces a jump from the ground (lowest) to the first excited quantum state. Although approximate, each vibration in a molecule can be associated with motion in a particular group. A simple example is methanal, whose six vibrations involve the following motions. A model IR spectrum is given in Figure 3.5.

The main spectroscopic techniques to detect vibrations in molecules are based on the processes of infrared absorption and Raman scattering. While Raman scattering involves, excitation of a molecule by inelastic scattering with a photon (from a laser light source), infrared spectroscopy typically involves photon absorption, with the molecule excited to a higher vibrational energy level.[5]

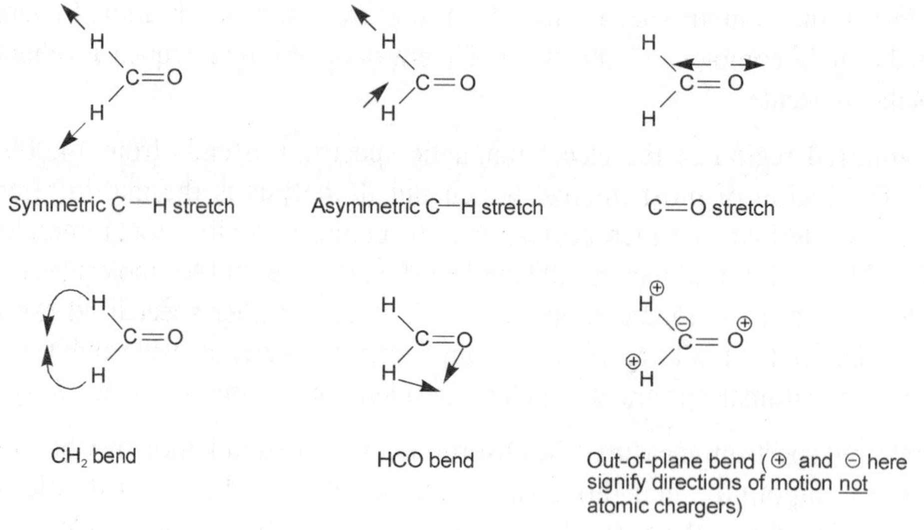

Figure 3.4 Vibrational modes of methanal

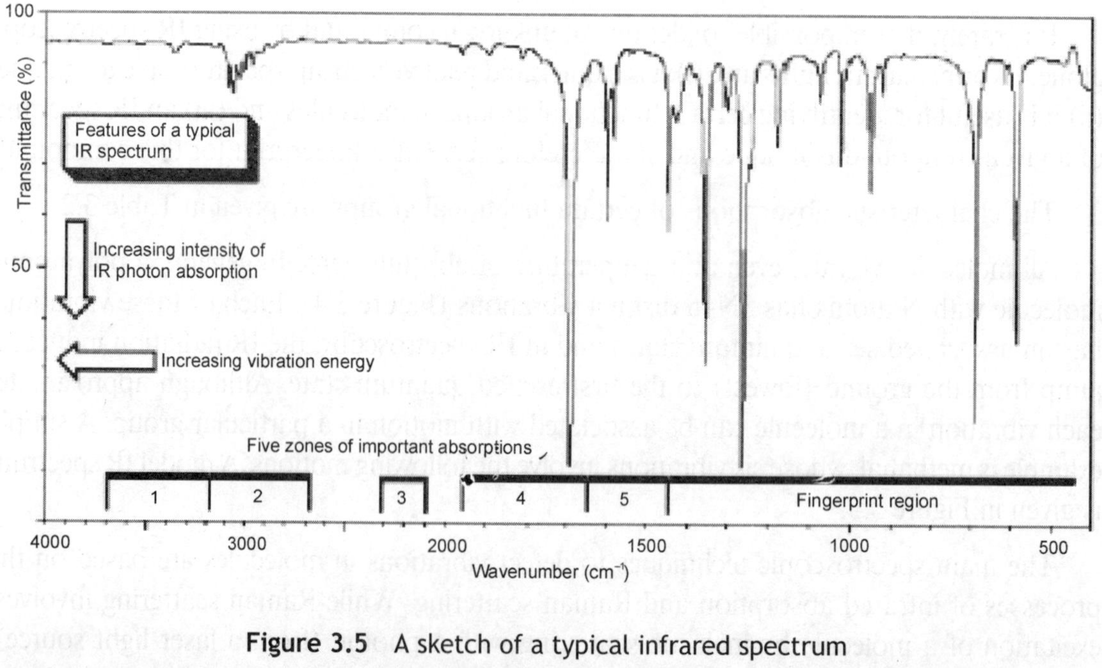

Figure 3.5 A sketch of a typical infrared spectrum

Table 3.2 lists the characteristic infrared absorption of some functional groups.

Table 3.2 Characteristic IR stretching frequencies

Functional group	Bond	Stretching, cm^{-1}	Intensity
Zone 1: 3700–3200 cm^{-1}			
Alcohol	O−H	3650–3200	Variable; usually broad
Alkyne	≡C−H	~3300	Strong
Amine, almide	N=H	3500–3300	Medium, often broad
Zone 2: 3200–2700 cm^{-1}			
Alkane	sp^3 C−H	2960–2850	Variable
Aryl, vinyl	sp^2 C−H	3100–3000	Variable
Aldehyde,	sp^2 C−H	~2900, ~2700	Medium, two bands,
Carboxylic acid	O−H	3000–2500	Strong, broad
Zone 3: 2300–2100 cm^{-1}			
Alkane	C≡C	2260–2100	Variable
Nitrile	C≡N	2260–2200	Variable
Zone 4: 1950–1650 cm^{-1}			
Aldehyde	C=O	1740–1720	Strong
Amidel	C=O	1690–1650	Strong
Aryl Ketone,	C=O	1700–1680	Strong
Carboxylic acid	C=O	1725–1700	Strong
Ester	C=O	1750–1735	Strong
Keton	C=O	1750–1705	Strong
Enone(C=C−C=O)	C=O	1685–1665	Strong
Aromatic overtone		1950–1750	Three or four small humps
Zone 5: 1680–1450 cm^{-1}			
Alkene	C=C	1680–1420	Variable
Aromatic	C=C	1600, 1500–1450	Variable; 1600 often two bands

Fingerprint region: <1450 cm^{-1}

Raman scattering depends on changes in the polarizability due to molecular vibrations, while infrared absorption depends on changes in the intrinsic dipole moments induced by molecular vibrations.[2]

Limitations IR is not a primary method. For exact identification, an unknown spectrum must be compared to a known spectrum on file. For quantitative analysis, measurements must first be made on compounds or mixtures whose composition has been determined by independent methods. Quantitative accuracy is typically 1%, although sometimes it cannot be done at all and at other times the accuracy may exceed 0.01%. Accuracy depends on the concentration of the structure or functional group, its absorptivity, and freedom of the band being used from overlap.

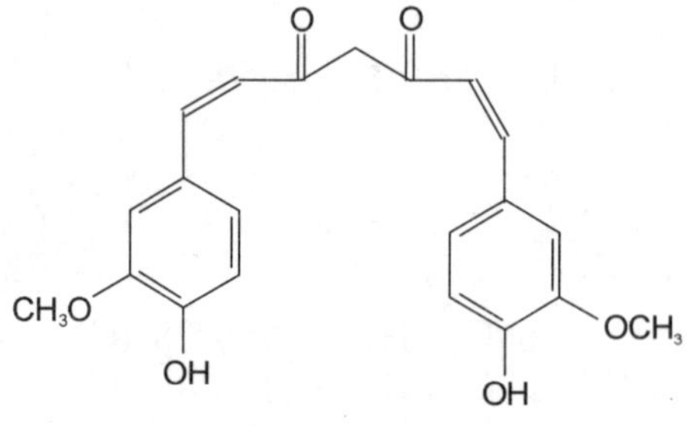

Chemical structure of curcumin

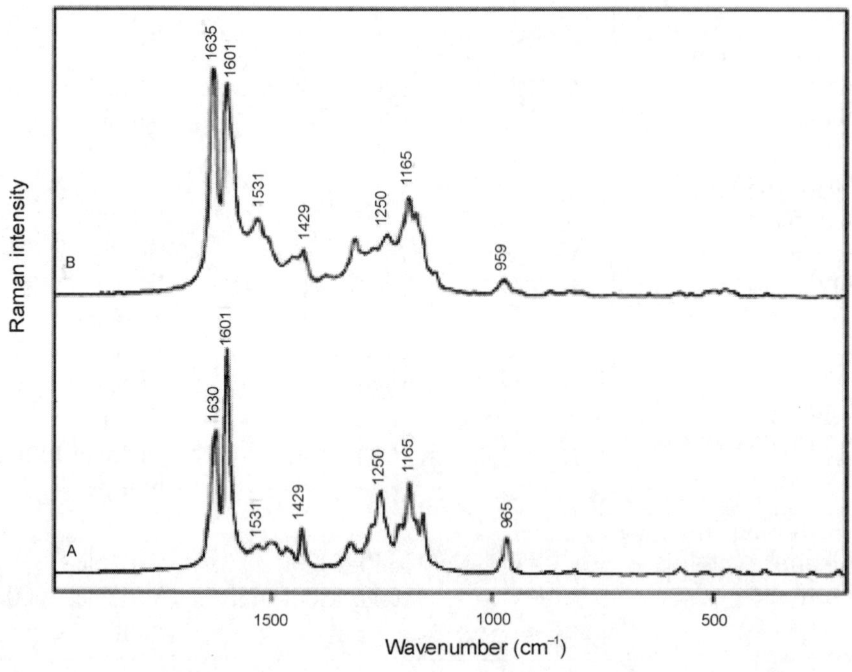

Figure 3.6 F T-Raman spectra of pure curcumin (A) and *Curcuma longa* root (B).

Raman and infrared spectroscopy thus provide "complementary" information about the molecular vibrations. While infrared (IR) spectroscopy has been established for decades as a useful tool for structure elucidation and quality control in various industrial applications, the use of Raman spectroscopy was restricted for a long time primarily to academic research. Nowadays, both spectroscopic methods are widely used to gain information on chemical structures, to identify substances from the characteristic spectral patterns ("fingerprinting") and to determine quantitatively or semi-quantitatively the amount of a substance in a sample.

Due to the strong dipole moment of water, which results in a strong signal, applications of IR spectroscopy were focused on dry material. Attenuated total reflectance (ATR) techniques have also evolved rapid IR measurements of many liquids and solvent extracts. Contrary to the strong absorption in IR, water has only weak Raman scattering properties making this technique more suitable to perform *in situ* studies of fresh plant material. Raman scattering is in general less widely used than infrared absorption, mainly due to problems with sample degradation and fluorescence. However, in the past years a renaissance of Raman spectroscopy was triggered by recent advancement in laser technology, by the design of very efficient filters to suppress elastically scattered Rayleigh light, by the development of extremely sensitive detectors and new methodological approaches for enhancing signal intensity. As an example, the chemical structure of curcumin (from turmeric) at its Raman spectra are given in Figure 3.6.

NUCLEAR MAGNETIC RESONANCE[7–12]

Complete structural elucidation of natural products often requires nuclear magnetic resonance (NMR) spectroscopy. NMR operates best with relatively pure sample solutions. However, the natural product of interest frequently resides in a complex matrix. Moreover, NMR possesses a lower sensitivity than other analytical techniques, thereby limiting its application to small-quantity samples. Unlike synthetic chemicals, natural products cannot always be procured in large amounts.

Structure elucidation of compounds isolated from plants, fungi, bacteria, or other organisms is a common problem in natural product chemistry. There are many useful spectroscopic methods to get information about chemical structures, mainly nuclear magnetic resonance (NMR) and mass spectroscopy. After the molecular formula has been determined (e.g., from a high-resolution mass spectrum), NMR spectroscopy assumes special importance, as it is the only method out of the four which achieves atomic resolution.

Today, NMR is recognized as the most all-round and informative spectroscopic technique available to a chemist.

NMR is the study of the magnetic properties (and energies) of nuclei. The absorption and emission of electromagnetic radiation can be observed when the nuclei are placed in a (strong) external magnetic field. NMR was developed around 1946 and also earned a Nobel Prize.

Atomic nuclei with non-zero spin adopt quantized orientations in magnetic fields.[3, 4] Transitions between these quantized states can be induced by radio frequency irradiation. The transition energies are modulated by the nature of the atom (i.e., ^{1}H, ^{13}C, ^{15}N, etc.) as well as by the electronic environment of each respective nucleus which is the origin of the different chemical shift values of atoms occurring in different molecular positions.

Since the energy differences between the quantized orientations of individual nuclei are small (in the range of μeV), the differences in the occupancy of the respective higher and lower level of a given nuclear species are small. This is the main reason for the relatively low-sensitivity of NMR spectroscopy; even with the most advanced instrumentation, at least 1 molar concentration of a given analyte is required. By comparison with other bioanalytical methods such as mass spectroscopy, gas chromatography, high-performance liquid chromatography and capillary electrophoresis, the sensitivity of NMR analysis is lower by orders of magnitude. However, this relatively low sensitivity is in part compensated by the abundance of molecular information that can be obtained from typical NMR spectra.

At this stage, NMR was purely an experiment for physicists to determine the nuclear magnetic moments of nuclei. NMR could only develop into one of the most versatile forms of spectroscopy after the discovery that nuclei within the same molecule absorb energy at different resonance frequencies. These so-called chemical shift effects, which are directly related to the chemical environment of the nuclei, were first observed in 1950.

NMR spectroscopy looks at how particles with spin behave in a magnetic field. To be NMR-active, i.e., able to generate NMR signals, spin has to be non-zero. Some nuclei are not detectable in NMR, e.g., ^{12}C and ^{16}O. Fortunately, spin = ½ for ^{1}H and ^{13}C, i.e., it allows NMR use these most abundant nuclei for structure determination of organic compounds.

The NMR resonance frequencies, or chemical shifts, give direct information about the chemical environment of nuclei, thereby greatly aiding in the unambiguous detection and assignment of compounds.

The interpretation of these spectra normally requires specialists with detailed spectroscopic knowledge and experience in natural products chemistry. NMR spectroscopy is the best method for complete structure elucidation (including stereochemistry) of non-crystalline samples. For structure elucidation of secondary natural products (not proteins), only ^{1}H NMR and ^{13}C NMR spectroscopy, including combined methods such as 2D NMR spectroscopy, are important because hydrogen and carbon are the most abundant atoms in natural products.

The Larmor frequency of a nucleus with a magnetic moment depends characteristically on the chemical environment of the nucleus and is expressed in terms of its chemical shift (δ)

$$\delta = \frac{\Delta \nu \cdot 10^6}{\nu}$$

where, $\Delta\nu$[Hz] is the frequency difference between the standard signal and compound signal. Only nuclei having a magnetic moment are investigated by nuclear magnetic resonance spectroscopy. Very important isotopes of organic chemistry, like ^{16}O or ^{12}C have no magnetic moment and therefore cannot be investigated with nuclear magnetic resonance spectroscopy. ^{1}H, ^{13}C, ^{14}N and ^{31}P are therefore the most important nuclei in natural products chemistry investigated by NMR.

^{1}H NMR and ^{13}C NMR spectroscopy are quite different. In a ^{13}C NMR spectrum every carbon atom occurs as a separate signal in most cases, while in ^{1}H NMR spectra many signals overlap and are therefore difficult to interpret. ^{1}H NMR spectra are logically decomposable, and the specialist can get direct information about the structure (including stereochemistry) from the resonance frequency and shape of the signals, provided a sufficiently high resolution of resonance signals can be experimentally achieved. In ordinary ^{13}C NMR spectra, only the resonance frequency is observed, and no signal splitting (decomposition) is possible. However, the absolute value of the resonance frequency provides more information about chemical structure and (important for prediction and database maintenance purposes) is not as sensitive to varying experimental conditions (such as the magnetic strength of the NMR spectrometer or the type of solvent) as ^{1}H NMR spectroscopy is. Additional measurements can be used to determine the number of hydrogens directly connected to a particular carbon atom. This number is the so-called multiplicity of the signal—S stands for singlet, which means there is no proton (i.e., hydrogen) connected to the carbon. "d" stands for a doublet (signal or peak splitting) with one proton connected to the carbon. " t" stands for a triplet with two protons and "q" for a quartet with three protons bound to the carbon atom.

Why ^{13}C NMR[13, 14]

First, a very general feature of ^{13}C spectra is their high spectral dispersion. Unlike ^{1}H spectra, very few signal overlaps occur in ^{13}C spectra even for molecules with several dozens of carbons. In many cases, merely counting the number of carbon signals in a (proton-decoupled) 1D ^{13}C spectrum can provide valuable information about the molecule under study. The reason for this is the large chemical shift range comprising more than 200 ppm for most organic molecules, in combination with the narrow lineshapes down to less than 0.01 ppm (i.e., < 1 Hz) because of slow relaxation. Usually there are distinct chemical shift regions for certain types of carbon atoms (such as carbonyls, aromatic carbons, aliphatic methyl groups, etc.) that allow for a quick orientation and rough assignment by the experienced spectroscopist.

This ease of interpretation is also the reason for the preference of most databases (i.e., CSEARCH and SPECInfo) for ^{13}C spectra.

This CSEARCH ^{13}C data bank; Biorad, and SPECInfo data bank with ^{13}C, ^{31}P, ^{15}N, ^{19}F, and ^{17}O NMR data; Chemical Concepts, Weinheim, Germany, 1989.

The information content of a 1D carbon spectrum can be easily stored by just providing the ppm values (and possibly multiplicities, corresponding to the number of directly bound protons) of each signal. By contrast, a proton spectrum would have to be stored as a complex ensemble of overlapping lines with different splitting. Furthermore, carbon is the only heteronucleus that is practically ubiquitous in organic compounds (comparable to ^{1}H). Other heteroatoms such as ^{15}N, ^{19}F, and ^{31}P occur only sporadically in organic structures, except for very special substance classes. Therefore, the only stable and NMR-active carbon isotope, ^{13}C, is the favourite NMR probe for carbon skeletons. Modern ^{13}C NMR spectroscopy can also provide a wealth of parameters that can be used for conformational and dynamic studies. Among these are the direct and long-range ^{1}H–^{13}C coupling constants $^{i}J_{H,C}$ with i = 1–3 that contain information about dihedral angles, as well as ^{13}C-T_1 relaxation rates for molecular mobility.

Types of Information from ^{13}C NMR Spectra

1. Splitting patterns in ^{13}C NMR are not important; all peaks are singlets; the total number of peaks, however, is important.
2. ^{13}C NMR signals cannot be integrated reliably; thus, no information can be obtained on the relative numbers of carbons in a group.
3. The chemical shift (δ, in ppm) gives a clue as to the type of carbon generating the peak (alkane, alkene, benzene, aldehyde, etc.). The chemical shift range is very large for ^{13}C NMR, approx. 220 ppm.
4. A separate experiment can be done to determine the number of Hydrogens directly attached to a carbon; this info is often indicated on the spectrum (q = 3H's, t = 2H's, d = 1H, s =0H)

One of the simplest NMR parameters is concurrently one of the most useful: the ^{13}C chemical shift describes with only one number, the complex chemical environment of a carbon atom. The chemical shift values for all carbon atoms of an organic substance can be determined easily and, yields a characteristic fingerprint of an unknown compound. This fingerprint is unique and theoretically sufficient to elucidate the structure of molecules with 15–20 non-hydrogen atoms, even if an experimental uncertainty of 0.5 ppm is assumed. For a higher number of non-hydrogen atoms, the number of possible constitutions increases faster than the number of possible different ^{13}C NMR spectra (within the experimental uncertainty). Consequently, two substances can yield quasi-identical spectra.

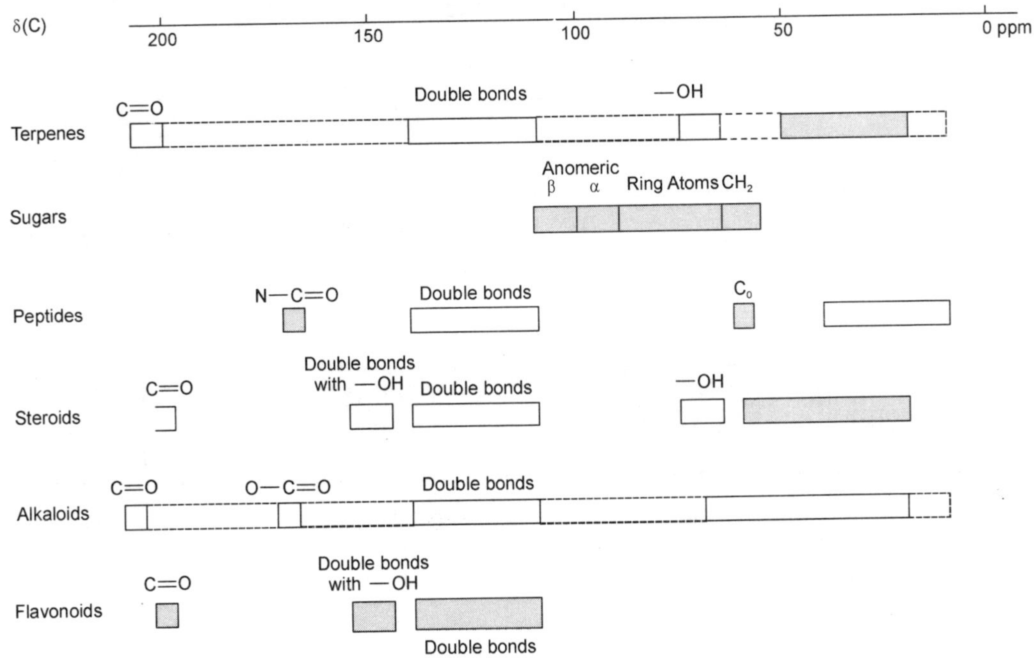

Figure 3.7 Typical ^{13}C-NMR signals of some commonly encountered natural products. Signals characteristic to the class indicated by filled bars, while additional signals are shown as empty bars.

NMR Spectra of Taxol

NMR spectra of Taxol is given in Figure 3.8.

^{13}C NMR Spectra of Terpenes and Steroids

The ^{13}C chemical shifts of organic molecules are spread over a range of 200 ppm. With increasing number of (20) hydrogen atoms attached to carbon atoms of hydrocarbons, the ^{13}C signal shifts generally to higher field.

Electron-withdrawing functional groups or heteroatoms cause downfield shifts especially on the resonances of neighbouring carbon atoms. ^{13}C chemical shift ranges of the main types of carbon nuclei are surveyed in Figure 3.7.

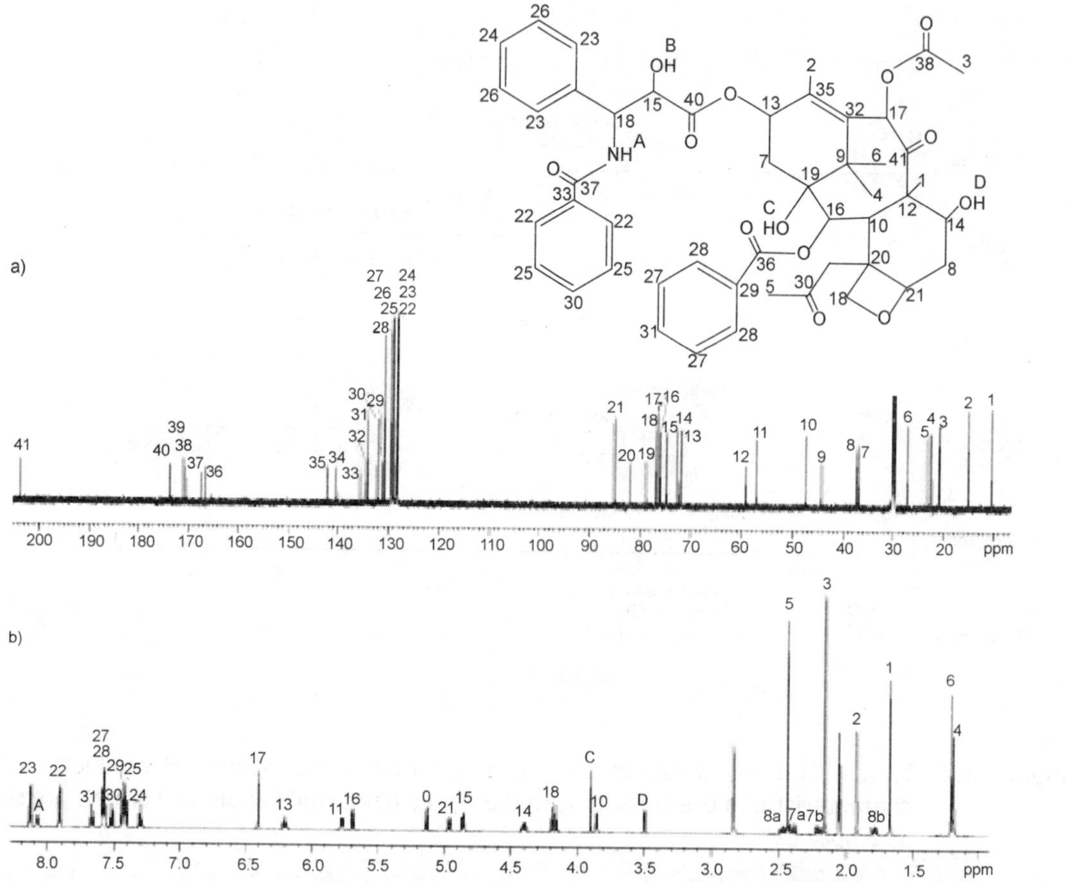

Figure 3.8 ^{13}C NMR and ^{1}H NMR spectrum of Taxol recorded in (D$_6$) acetone at 150 and 600 MHz

As a demonstrating example for a ^{13}C NMR spectrum of higher molecular weight terpenes and steroids, that of nimbin is given (Figure 3.9). The signal assignments are done on the following basis:

1. application of general chemical shift rules
2. determination of the number of hydrogen atoms attached to each carbon atom by proton off-resonance spectroscopy,
3. spectral comparison with similar and partial structures.

Polyols are reduction products of carbohydrates and often used for identification of the parent compound. The ^{13}C NMR spectra of polyols are much more easy to interpret than those of their corresponding sugars because they cannot undergo mutarotation.

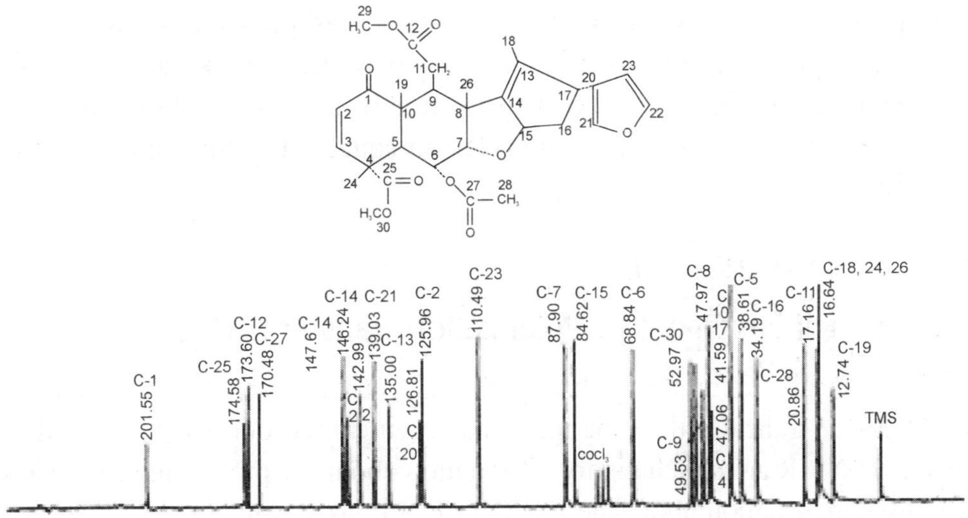

Figure 3.9 22.63 MHz PFT ^{13}C NMR spectra of nimbin, 350 mg/1.5 ml $CDCl_3$, temperature: 30°C.

10 p sec, pulse interval: 0.4 sec/4k interferogram; phase corrected. (a) proton off-resonance decoupled; 9550 accumulated interferograms; (b) proton broad band decoupled; 1600 accumulated interferograms. Figure 3.10 shows the NMR spectrum of ribitol.

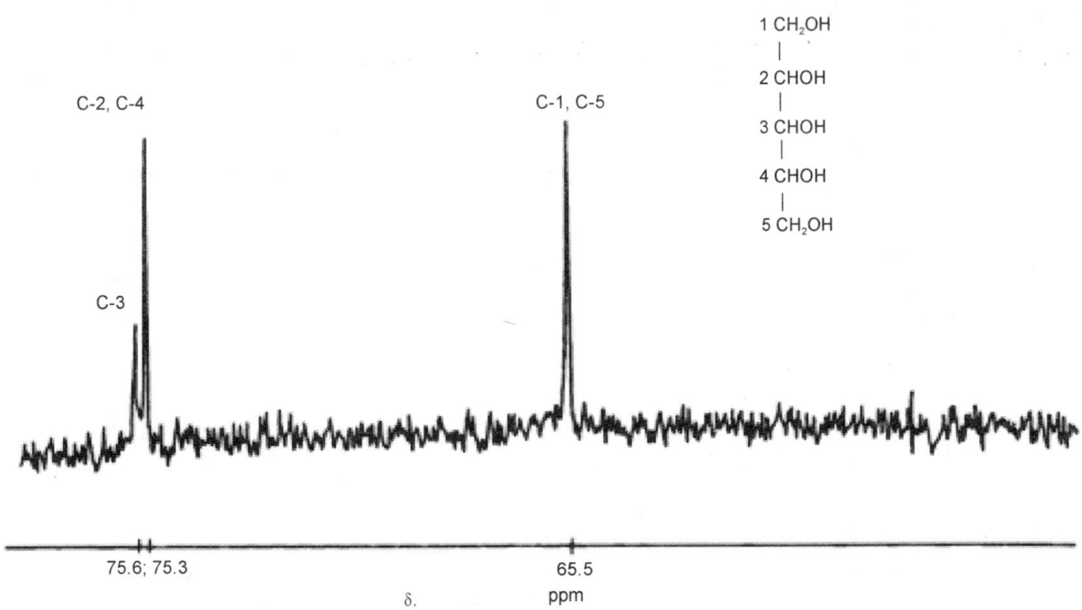

Figure 3.10 22.63 MHzPFT $^{13}C\{°H\}$NMR spectrum of ribitol, 20% in D20, temperature: 30°C,pulse width: 10 sec, pulse interval: 0.4 sec/4 Kinterferogram, ppm values are given relative to TMS = 0.

CMR spectroscopy gives information about the carbon skeleton of an unknown compound. Compared to almost all other spectroscopic methods it is possible to characterize all the information of a CMR spectrum by a small set of digital values, and the number of signals of a CMR spectrum is less or equal to the number of carbon atoms of the molecule.

ADVANCED NMR TECHNIQUES

Distortionless Enhancement by Polarization Transfer (DEPT)

The Dept spectrum gives information about CH_3, CH_2 and CH-carbons (not about quartenary carbons) Dept-135 indicates the experiment has been carried out using a 135° decoupler pulse. While ^{1}H NMR gives preliminary information about the proton and its environment ^{13}C NMR gives information about the number of carbons and their environment to some extent. Dept-adds information about the methyl, methylene and methine carbons. Figure 3.11 gives a snapshot of the DEPT-spectra of IPsenol. Table 3.3 shows the various NMR experiments and their information.

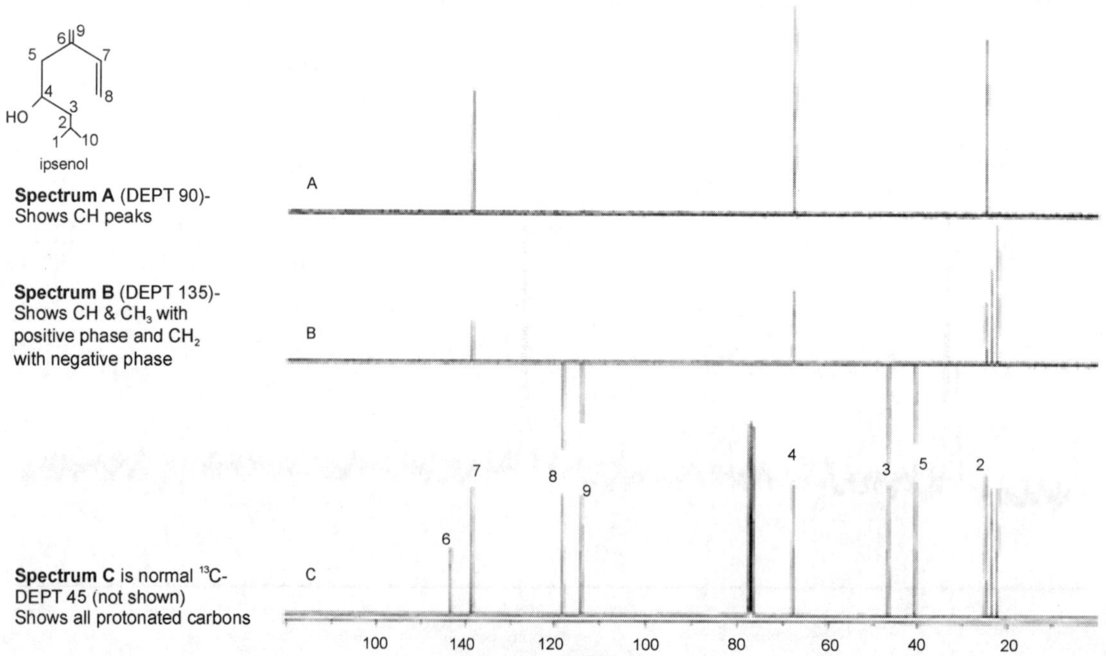

Figure 3.11 The DEPT spectra of IP senol

Table 3.3 Various NMR experiments and their information

NMR experiment	Information
H	Identification of kind of protons, number of non-exchangeable protons.
COSY	Identification of protons close together (separated by 2 or 3 bonds)
TOCSY	Identification of spin systems
HSQC or HMQC	Identification of connected proton-carbon, CH_2 identification, what kind of carbon.
HMBC	Identification of connected proton-carbon by 2 or 3 bonds, carbons without attached proton.

Nuclear Overhausser Effect

Nuclear Overhausser effect arises from dipole–dipole interactions of atoms in close proximity to one another (up to 5 Å). NOESY (Nuclear Overhauser Effect Spectroscopy) spectra provide information about protons that are 5 Å or less apart in space. The information is through space and not through bond, like a COSY or TOCSY. The absence of an NOE peak between protons does not necessarily mean that they are not within 5 Å since other factors can reduce a NOE peak even if the protons are close in space. A midsize molecule (~1000–1500 MW range) may have NOEs that are close to zero and a ROESY may be required to see them. Large molecules generally give better NOEs at higher field, but small molecules may actually give better NOEs at lower field. A 2D NOESY of a small molecule will have cross peaks of opposite phase to the diagonal. A 2D NOESY of a large molecule will have cross peaks of the same phase as the diagonal. This is distinct from COSY-type spectra which use the J-coupling interaction to report on which proton resonances are located on the same or adjacent carbon nuclei.

NOE interactions can be very useful in determining internuclear separation by irradiating a particular resonance, atoms within range will be affected by either an enhancement or depression of the signal. Subtracting the reference spectrum from the enhanced spectrum reveals the resonances experiencing NOEs. One-dimensional NOE detects steady-state NOEs and therefore is applicable only to MW < 1000.

One of the main practical differences between NOESY spectra and COSY/TOCSY spectra is the fact that the NOE interaction is relatively weak. This results in NOESY spectra having much lower-intensity cross-peaks than one would expect from a COSY/TOCSY spectrum of the same sample, even if the NOESY spectrum is a "good" spectrum with no obvious problems. This results in a situation where there is not the same type of linear relationship between the "results" of a NOESY spectrum (usually considered to be the overall signal intensities of the cross-peaks) and the quality of the sample and experimental set-up as there is for COSY-type spectra. In other words, if one runs a COSY experiment on a decent sample, barring instrument problems or gross errors in experiment set-up, one can reasonably expect to see cross-peaks of decent intensity in the resulting spectrum. However, this is not always the case with NOESY spectra.

Figure 3.12 and 3.13 represent the heteroCosy and NOE spectrum of a 3, 5-dimethyl pyran2-one and phytochemical respectively.

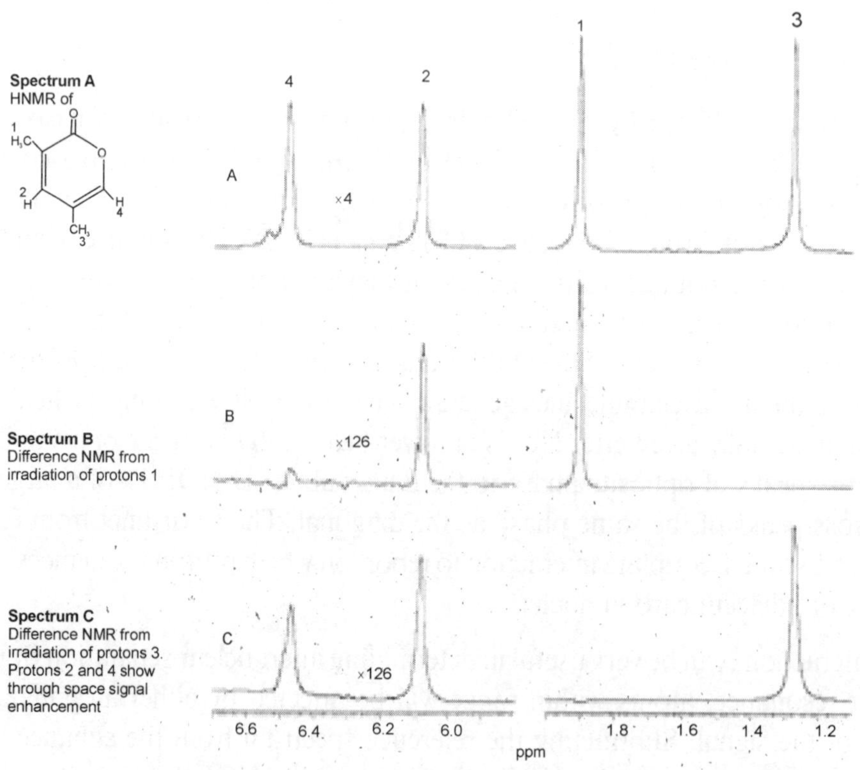

Figure 3.12 The NOE spectrum of 3, 5-dimethylpyran-2-one

Polyhydroxylated cyclic 13-membered sulphoxide isolated from aqueous extract of Kothala-himbutu (*Salacia reticulata*).

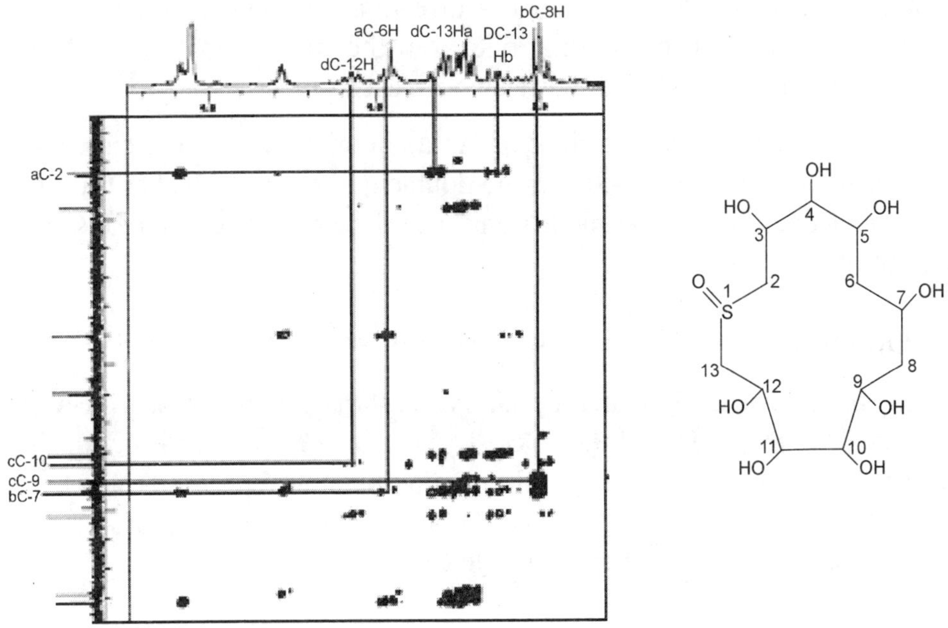

Figure 3.13 The HeteroCosy spectrum of a phytochemical

2DCOSY (CORRELATION SPECTROSCOPY)

Figure 3.14 gives the homocosy spectrum of a terpene. As the name indicates it is a x-y plot of ^{13}C spectrum.

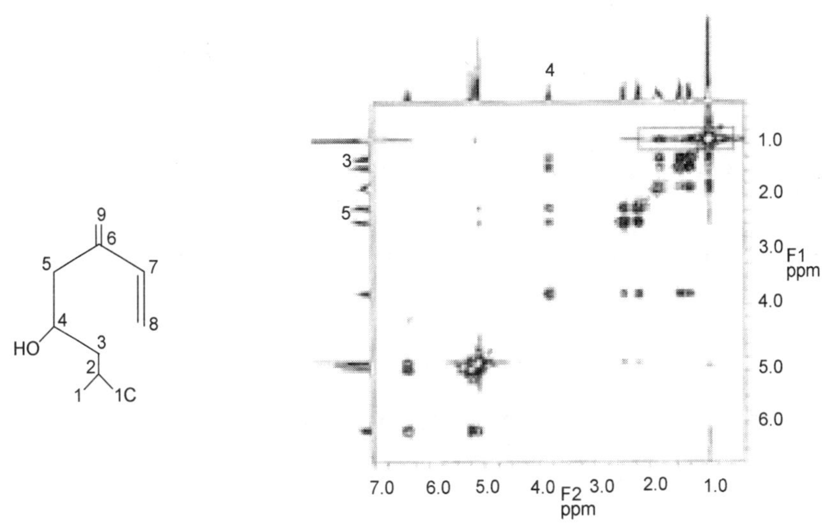

Figure 3.14 The HomoCosy spectrum of a terpene

Leaving the points in the diagonal (self-correlation), if a line is drawn from a carbon peak from y axis and a corresponding one from the from x-axis, it results in a peak (smudge 1). This indicates that they correlate with each other.

This is a modified 2DCOSY technique. An additional pulse to pulse sequence, single quantum transitions are suppressed, leaving double quantum and higher transitions. As a result, noncoupled resonances like methyl singlets, will be greatly reduced.This improves the resolution of the spectrum.

Double Quantum Filtered Correlated Spectroscopy (2DQF-COSY)

DQF-COSY is the protype 2DNMR that identifies scalar, through-bond couplings between protons. Couplings should be resolved to give rise to cross-peaks. One ^{1}H spectrum is shown on each axis. While analysing, start from cross-peaks on the diagonal and trace either vertical or horizontal off diagonal cross-peaks indicate coupled protons. The number of protons in a particular spin system is readily determined. Figure 3.15 shows the DQF-COSY spectrum of terpene.

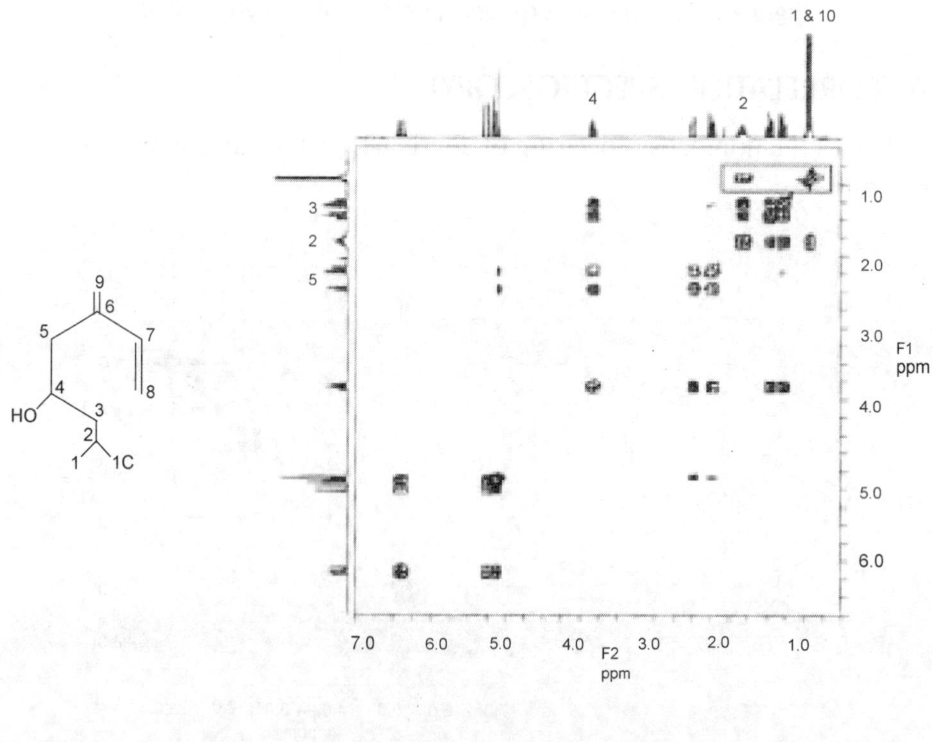

Figure 3.15 DQF-COSY spectrum of the terpene

TOCSY (TOTAL CORRELATION SPECTROSCOPY)

TOCSY or the HOHAHA (HOmonuclear HArtmann HAhn) Spectroscopy shows the correlation between protons in the same spin system. Table 3.4 shows the NMR methods, keywords and its references.

Table 3.4 NMR methods and their keywords and references

COLOC	COrrelation via LOng range Coupling
1D	One dimensional
2D	Two dimensional
DEPT	Distortionless Enhancement by Polarization Transfer
HMBC	Heteronuclear Multiple Bond Correlation
HMQC	Heteronuclear Multiple Quantum Coherence
INADEQUATE	Incredible Natural Abundance Double Quantum Transfer Experiment
INEPT	Insensitive Nuclei Enhanced by Polarization Transfer
J-J-Coupling, Scalar Coupling	
JBCA	J Based Configuration Analysis
ROESY	ROtating frame nuclear Overhausser Effect Spectroscopy

Computer-aided Interpretation of the NMR Spectra

A number of powerful aids for NMR spectroscopic interpretation[15–18] have emerged and are now readily available from instrument manufacturers and include programs such as Auralia and AMIX.

However, any comprehensive CASE program will be based on a quality structure generator. To this point, only a few high-quality, pure structure generators have been developed. While over the years there have been a variety of programs that have been developed, they have been limited to those researched and developed by small, private groups. Now there are some highly capable programs that are commercially available, such as Assemble (Upstream Solutions GmbH, Zurich, Switzerland) and MOLGEN (http://www.molgen.de). A wellknown deterministic CASE system that is often cited in the literature is CISOC-SES, which is now commercially available under the acronym NMRSAMS (Spectrum Research, Madison, WI, USA). Another deterministic CASE program, COCON has been relatively recently introduced, and several examples demonstrating its value in structural elucidation have been reported.

Deterministic algorithms have a limit to the size of the molecule with which they can work. To overcome this, a stochastic structure generator has been published for the computer-assisted structure elucidation of organic molecules.

The name of the program is SENECA. This program is written in Java and therefore is platform independent and allows for a simple plug-in mechanism for new spectroscopic data types. Theoretically, many different kinds of properties can be plugged into this system, as long as the property can be reliably calculated from a generated molecule. Classical CASE systems can at least attempt to provide the two-dimensional structure of an unknown molecule. Now, with the greater exposure of, availability of, capabilities of, and demand for CASE programs, efforts are being made toward incorporation of the ability to confidently determine the three-dimensional structure of unknown molecules.

Advancements in chromatography, spectrometry, and spectroscopy together with breakthroughs in the coupling of these technologies are important steps in the production of a fully automated and integrated natural products structure determination instrument, which will provide significant advantage to the early, rapid, and facile identification of new natural-product-based drug opportunities.

DATA PROCESSING AND ANALYSIS

- **matNMR** matNMR is a highly flexible processing toolbox for NMR and EPR spectra under MATLAB (all platforms). It can read in binary FIDs from most commercial spectrometers, process any 1D or 2D spectrum and uses all graphical capabilities that MATLAB has to offer. The program is fully GUI and at all times allows the user to profit from MATLAB's numerical powers (Courtesy of Jacco van Beek, ETH Zurich).
- **NMRPipe** NMRPipe provides comprehensive facilities for Fourier processing of spectra in one to four dimensions, as well as a variety of facilities for spectral display and analysis. It is currently used in over 300 academic and commercial laboratories (courtesy of the Ad Bax group).

- **MestRe-C** from Spain - MestRe-C 2.0 is a program designed to process and plot 1D NMR spectra on PCs running Windows 95/98/NT 4.0. It imports files from most spectrometers and includes a number of useful tools (Courtesy F. Javier Sardina, Universidad de Santiago de Compostela, SPAIN).
- **RMN** A Nuclear Magnetic Resonance (NMR) data processing program for the Macintosh. Dr.Philip J. Grandinetti.
- **Gifa** "The Gifa program is a multipurpose NMR program. It is designed for the processing, display and analysis of 1D, 2D & 3D NMR data-sets." M.A.Delsuc, CBS, France.
- **MARDIGRAS** "(Matrix Analysis of Relaxation for DIscerning the Geometry of an Aqueous Structure) is a FORTRAN program for calculating proton–proton distances from cross-peak intensities measured from a 2D NOE experiment."—UCSF, USA.
- **Sparky** Graphical NMR assignment program for proteins and nucleic acids. Runs under Windows, Linux, or Unix.
- **Azara** Azara is a suite of programs to process and view NMR data. Courtesy Wayne Boucher.
- **iNMR (evolution of SwaN-MR)** "An NMR processing software for the Macintosh. It reads 1D, 2D, 3D and 4D spectra coming from many spectrometers and its feature list is comparable to those of commercial packages". Giuseppe Balacco.
- **CARA** CARA is a program for computer-aided resonance assignment of multidimensional NMR spectra of macromolecules. Dedicated tools along with a well-organized database allow for efficient assignment. Custom applications can be developed with scripts. All major platforms are supported.
- **MARS** MARS is a program for robust automatic backbone assignment of $^{13}C/^{15}N$ labelled proteins (courtesy Markus Zweckstetter, Max Planck Institute for Biophysical Chemistry, Germany).
- **HiRes** High Resolution Spectroscopy (HiRes), is a free Windows software program to provide comprehensive analysis of large NMR spectroscopic data sets. HiRes combines standard spectral processing routines, data correction functions, techniques for reducing information complexity of multi-spectral dataset such as Principal Component Analysis (PCA), and tools for Pattern Recognition, such as Non-Negative Matrix Factorization (NMF), which extracts a set of spectral patterns with direct physical interpretation that can be used for identifying meaningful biochemical effects. These spectral patterns are also correlated with auxiliary

information from the samples, such as biological end-points, drug toxicity or disease diagnosis. (Courtesy of Columbia University, New York, USA).

- **3DiCSI** 3D Interactive Chemical Shift Imaging (3DiCSI) is an interactive, user-friendly, and comprehensive software program for multi-dimensional CSI data visualization, spectral processing/analysis, spectral localization, and quantification. It currently runs on Windows platform and is provided as a software free for research. Courtesy of Qi Zhao, Columbia University.

COMMERCIAL PRODUCTS

- **Bio-Rad Laboratories, Inc., Informatics Division's—HaveItAllTM NMR**. This new product offers spectroscopists and other analytical chemists a reliable source of NMR data of over 140,000 ^{13}C and 12,000 ^{1}H spectra they can use as a reference in the first, fully integrated environment for NMR. The HaveItAll NMR database works within and includes Bio-Rad's new KnowItAllT Analytical System. With the combined power of the high-quality HaveItAll NMR data and the KnowItAll environment, researchers can have it all and know it all in one place-make predictions, search, access reference spectra, build databases with assignments (with KnowItAll's Database Building Option and AssignIt(TM) Option), and even cross-reference NMR data with other analytical techniques, such as IR, UV/Vis, GC, MS, near IR, and Raman, and generate high-quality reports.
- **Chenomx Inc.—Chenomx NMR Suite** Specialized NMR spectra analysis platform for metabolic profiling. By integrating a broad range of tools into the platform, Chenomx NMR Suite delivers unparalleled power and ease of use to assist you with NMR spectra analysis and interpretation.
    - **Chenomx Profiler** Assists you in identifying and quantifying metabolic compounds within bio-fluid samples. Using a Compound Library of over 200 unique spectral signatures together with computer-assisted routines, a user can accomplish in minutes what would have taken hours or days using manual analysis methods;
    - **Chenomx Signature Builder** allows users to create signatures that model the compounds of their particular interest;
    - **Chenomx Library Manager** enables users to construct customized compound library for specified projects;
    - **Chenomx Processor** allows users to convert a variety of native spectrum formats, such as Varian, Bruker, and JCAMP into the format compatible with Chenomx NMR suite.

Seeing is believing—please visit http://www.chenomx.com to download and try the evaluation version of Chenomx NMR Suite 4.0.

- **magritek's PROSPA** NMR Processing Package for Windows especially suitable for NMR Imaging and Research applications. Version 1.9 has just been released. Developed by Dr. Craig Eccles. New features coming soon include 2D Laplace inversion and enhanced 3D visualisation. Can process and display 1D, 2D and 3D data sets acquired with a number of different NMR spectrometers. http://www.magritek.com/prospa
- **Spectrum Research** CASE As a part of a new generation of "Computer Assisted Structure Elucidation (CASE)" programs we have developed an expert system of programs called SpecMan and NMR-SAMS, which take into account the spectroscopist's experience and knowledge into the process of spectrum interpretation and structure generation. They provide the spectroscopist the opportunity to work more efficiently and rapidly.
- **ScienceSoft's NMRanalyst** Computerized analysis of phase-sensitive 1D through 3D NMR FIDs and spectra. Generates accurate spin system descriptions including signal integrals and coupling constants. Sold by Varian, Inc. as "Full Reduction of Entire Datasets", FRED(tm).
- **Umatek's ADVASP** The ADVASP spectra analysis software is designed especially for total customer satisfaction in analysing IR, MASS and NMR spectra. The main idea in creation of this program is to make user interface as easy as possible from one side and as functional as it is required by professionals.
- **ACD/Labs' NMR Prediction Software** — ACD/HNMR enables you to calculate the proton NMR spectrum for any organic structure to a high accuracy. Prediction is based on an internal data file with over 1,384,000 experimental chemical shifts and 449,000 coupling constants. ACD/CNMR brings you the fastest and most accurate 13C NMR prediction engine available based on the analysis and correlation of over 2,017,000 observed chemical shifts and 81,000 coupling constants. ACD/Labs also offers NMR prediction modules for 15N, 19F, and 31P.
- **Woody Conover's (Acorn)** SAM, NUTS, and Virtual Spectrometer.
- **Accelrys'** FELIX-NMR data processing, analysis and assignment program.

STRUCTURAL ELUCIDATION THROUGH THE WEB

The structural elucidation of natural products is a great challenge due to its great structural diversity and complexity. For this reason, a database with information on ^{13}C NMR spectra, accessible through a standard browser, is being developed (http://c13.usal.es). Currently it contains the structures of several thousands of compounds, along with their ^{13}C NMR information with a continuous increasing, because this database has capacity for storing hundreds of thousands of compounds. The utility of the database will be conditioned on the number of entered compounds. The database has many search facilities and an appearance that allows the comparative study of related compounds. At present, new search tools are being developed and the data input methods are being improved in order to allow researchers from different institutions to enter the information through the Net.

The aim of this database is to help identifying or elucidating the structure of a hypothetical new compound, by comparing its ^{13}C NMR data with those related already published. Several tools that will facilitate this task have been developed—search by substructure in a graphics environment, search by chemical shift, both previous search methods combined, search refinements, results displayed in different layouts in order to make a comparative study, deviation calculus in fixed positions, etc. Also, we have developed scripts to automatically parse input data, run different tests and populate the database.

MASS SPECTROMETRY[19-23]

Mass Spectrometry (MS) was primarily used to obtain molecular weights. Former ionization techniques, such as electron ionization, limited this use to non polar, volatiles and thermostable substances. The improvement of softest ionization techniques allowed gradually the analysis of polar and thermolabile compounds, having currently unlimited physical or chemical properties to be analysed by MS.

But the development of tandem MS allowed MS could be applied also to structure elucidation. Furthermore, hyphenated MS techniques, such as GC, LC and CE MS, permit the analysis of very low sample amounts and no requirement of purity.

Mass spectrometry (MS) is a powerful analytical technique for identification of unknown compounds by determination of mass-to-charge *(m/z)* values of intact molecular ions and their fragments. Mass spectra, reflecting physical and chemical properties of the analytes, provide qualitative and quantitative data for structure determination of synthetic low and high molecular compounds and for the wide diversity of natural compounds.

Without any doubt mass spectrometry has become one of the most valuable and efficient tools in the structure elucidation of natural products and an enormous amount of work has been published in the past which is surveyed in a series of excellent monographs. Mass spectra upto 3539 can be recorded from very low sample quantities (${}^{1}O^{-6} - {}^{1}O^{-10}$ g) and a great deal of structural information is received from them.

Sufficient volatility of a compound, however, is necessary if the most common method, the electron bombardment mass spectrometry, is applied. Polar functional groups and high molecular weights of natural products often prevent recording their mass spectra with this method. The volatility, however, can be often enlarged by simple chemical modifications of the polar groups like methylation, trimethylsilylation or trifluoroacetylation.

Example

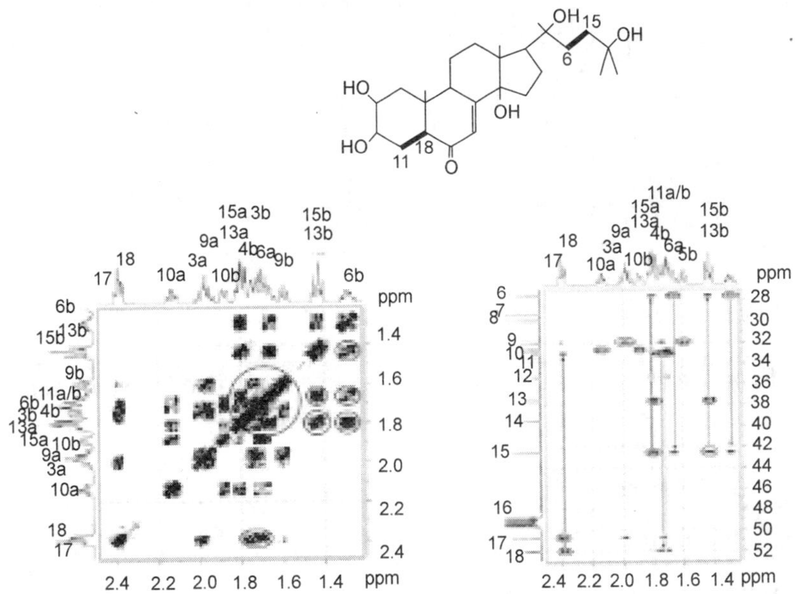

Figure 3.16 NMR analysis of the steroid b-ecdysone.Expansions of a) the DQF-COSY and b) the corresponding HSQC-TOCSY spectra. Encircled signals are HSQC peaks.The numbering does not correspond to the steroid numbering.

CASE STUDY A

ISOLATION AND STRUCTURE ELUCIDATION OF TWO COMPOUNDS FROM THE FRUIT PULP OF *MELIA AZADIRACHTA* LINN.[24, 25]

(A) Fruits from *Melia azadirachta* Linn. are extracted with ethanol (95%). The extracts are evaporated until green material deposits which is separated from the solution. After evaporation a green residue is received which is dissolved in ether, washed with water and dried over Na_2SO_4. From the solution crystalline needles are received which are chromatographically pure after recrystallization. The compound has a melting point of 166°C, an $[a]^{20}$ value of +11.2° and a molecular weight of 450 (on the basis of its mass spectrum). $C_{28}H_{34}0_5$ is calculated as molecular formula. An absorption at e =232.5 nm =25,228 gives evidence for an a,b -unsaturated keto chromophore to be part of the molecule. Figure A shows the IR spectrum of the compound registered from a KBr pellet.

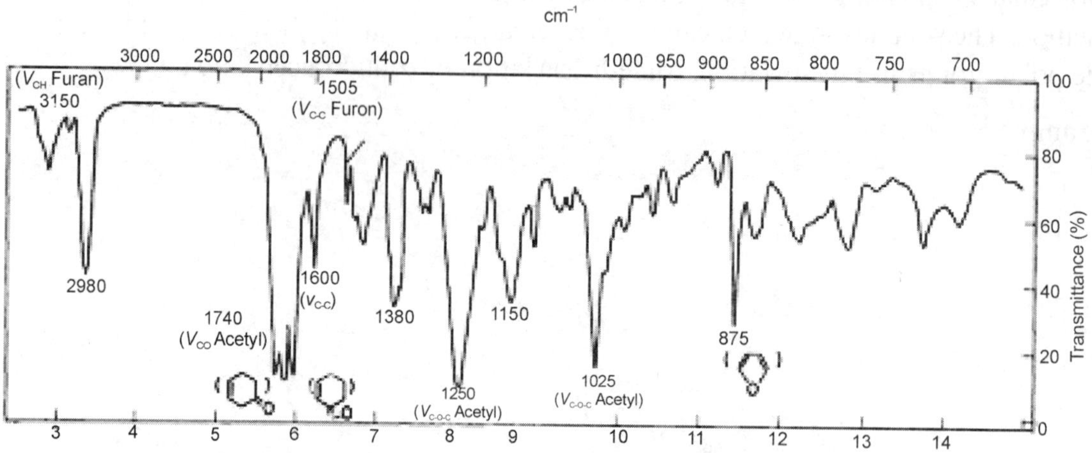

Figure A Infrared spectrum of the unknown natural product (1.5/30mg)

Functional groups	Band position [cm^{-1}]
Acetyl group	1025
Acetyl group	1250
Acetyl group	1740
α,β-Unsaturated cyclohexenone	1600
α,β-Unsaturated cyclohexenone	1665
Furan ring	875
Furan ring	3150

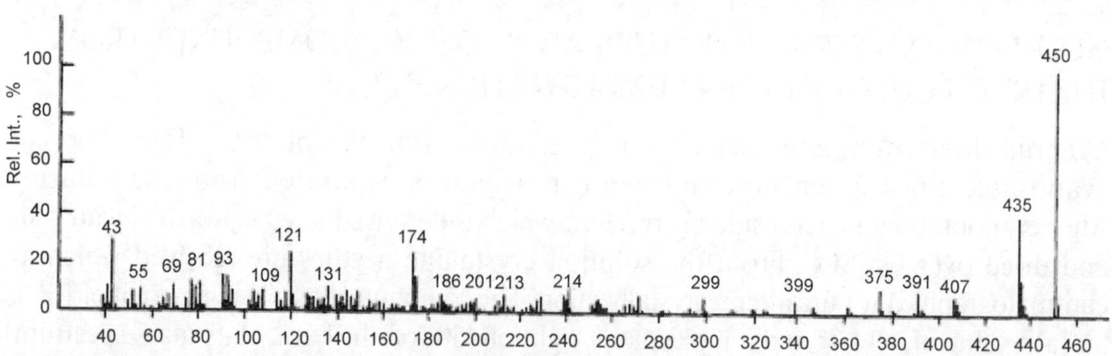

Figure B Mass spectrum of the natural product (LKB instrument, 70 eV, 3.5 kV).

According to the mass spectrum the molecular weight of the product is 450.

Interpretation of the most important ions in the mass spectrum of the natural product

m/e	Interpretation
450	M^+
435	M—CH_3
407	M—CH_3CO
391	M—CH_3COO
390	M—CH_3COOH
43	CH_3CO

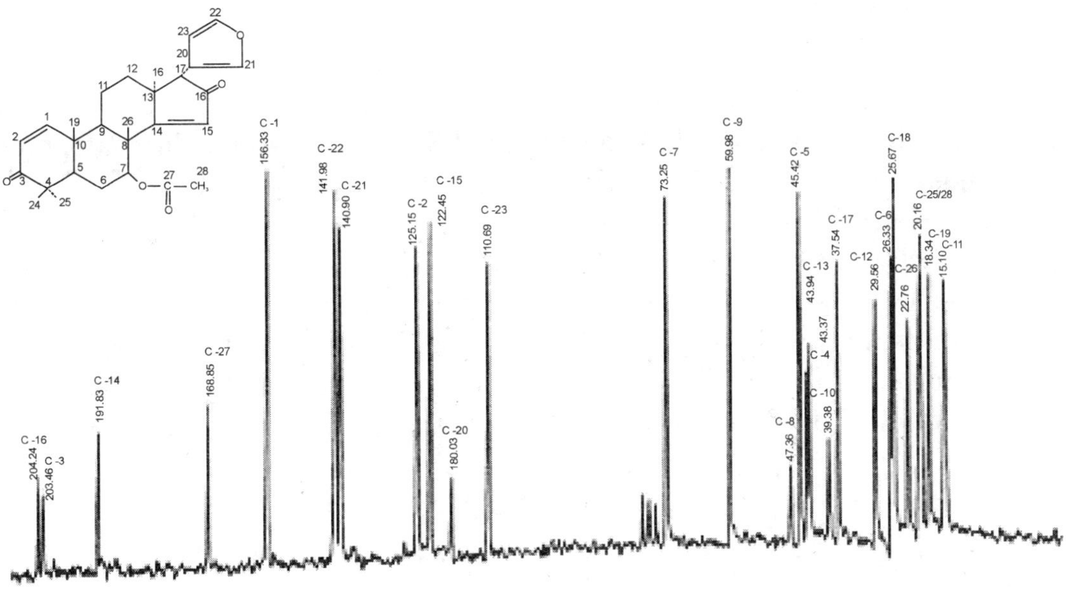

Figure C PFT ^{13}C NMR spectrum of the natural product, 22.63 MHz, 500 mg/1.5 ml $CDCl_3$, temperature: 30°C, pulse interval: 0.4 sec/4 K interferogram, phase corrected, accumulation of 2048 pulse interferograms, ppm values relative to TMS = 0

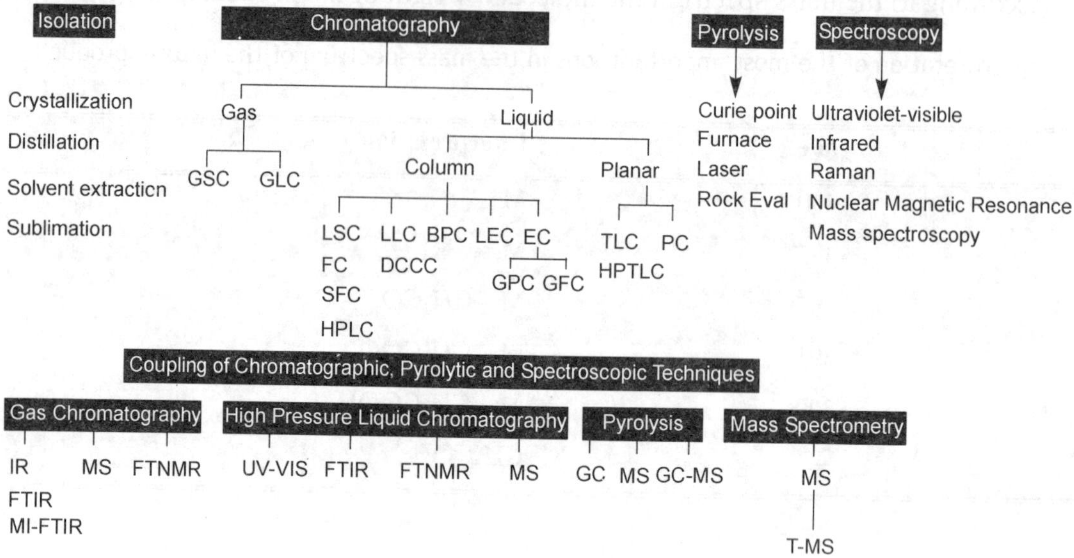

Figure D Chromatographic and spectroscopic techniques for detection and identification of organic compounds.[26]

GC—Gas chromatography, GLC—Gas–liquid chromatography, GSC—Gas–solid chromatography, TLC—Thin layer chromatography, HPTLC—High performance thin layer chromatography, PC—Paper chromatography, LSC—Liquid–solid chromatography, FC–Flash chromatography, SFC—Supercritical fluid chromatography, LLC—Liquid–liquid chromatography, DCCC—Droplet counter current chromatography, PBC—Bonded phase chromatography, HPLC—High pressure liquid chromatography, IEC—Ion exchange chromatography, EC—Exclusion chromatography, GPC—Gel permeation chromatography, GFC—Gel filtration chromatography, IR—Infrared, UV—Ultraviolet, NMR—Nuclear magnetic resonance, MS—Mass spectroscopy, FT—Fourier transform, T-MS—Tandem mass spectroscopy, MI-FTIR—Matrix isolation Fourier transform infrared.

CASE STUDY B

NAVIRINE B[27]

Navirine B is a diterpene alkaloid isolated from the aerial parts of *Aconitum naviculare*—a medicinal plant used in traditional Nepalese medicine. The compound had a molecular formula of $C_{30}H_{40}N_2O_3$ on the basis of the protonated molecular ion [M+H]+ displayed at *m/z* 477.3122 in the HRAPITOFMS. The IR spectrum showed absorption bands supporting the presence of an imino group (1670 cm^{-1}).

The ^{1}H NMR spectrum showed signals ascribable to one methyl group (δ 1.05 s, 3H), an *N,N*-dimethyl group (δ 2.40 s, 6H), and one olefinic proton (δ 5.67 brs, 1H). Two *ortho* coupled doublets (J = 8.0) in the aromatic region (δ 6.85 and 7.12, 2H each) also indicated the presence of an *o-p* disubstituted aromatic ring. The structure of compound, a diterpenoid alkaloid, was obtained by exhaustive analysis of COSY, HMQC and HMBC data. Various spin systems were detected when COSY and HMQC data were compared, establishing the connectivity between CH_2 in positions 1, 2, 3 and 6, 7 and between positions 9, 11, 12, 13. Further connectivity between the highly deshielded CH-19 (δ H 7.43 s, 1H; δ C 169.9) and CH-20 (δ H 3.57 brs, 1H; δ C 80.6) and also between the olefinic CH-15 (δ H 5.67 brs, 1H; δ C 130.8) and CH_2-17 (δ H 4.55 brs, 1H; δ C 68.8) were seen in the COSY spectrum. Diagnostic long-range (HMBC) correlations were observed from the methyl group 18 and carbon resonances at δ 30.9.

(C-3), 44.9 (C-4), 72.6 (C-5) and 169.9 (C-19). HMBC correlations, observed from the proton signal of CH-20 (δ 3.57) with carbon resonances C-19 and C-5, supported the presence of a six-member ring containing an imino group. Diagnostic long-range correlations were observed from the same proton signal (H-20) with C-1 (δ 27.9), C-9 (47.1), C-5 (72.6) and C-8 (43.8). HMBC correlations of the proton at δ 5.67 (H-15) with C-9 (δ 47.1), C-12 (δ 31.7) and C-17 (68.8) were also seen. These observations, as well as analysis of COSY and HMQC data, suggested two more hexa-atomic rings in the compound. Long-range correlations observed from proton signals at δ 1.91-1.56 (CH_2-13) with carbon resonances at δ 146.8 (C-16) and 80.6 (C-20) supported the linkage between positions 20 and 14.

Long-range correlations observed from the aromatic doublet δH 7.12 (H-3´ 5´) with the carbon at δC 33.3 (C-7´) and from the methyl group at δH 2.59 (CH_3-9´) with the carbon at δC 52.5 (C-8´) suggested the presence of one hordenine moiety in the molecule. NOESY data and comparisons with the standard spectra for navirine evidenced the relative stereochemistry of the molecule, supporting an α orientation for groups at positions 10, 8 and 12. NOESY cross-peaks were also obtained between the exchangeable proton signal at δ 3.48 (OH-5) and H-9 (δ 1.64), which were assigned as β on the basis of previous references[4–7], suggesting a β orientation for the hydroxy group. With all this evidence, compound 1 was established as a new alkaloid, called navirine B (5β-hydroxy navirine).

amorphous solid.

[α]D: +12.0 (c 0.083, CH_3OH).

IR (KBr): 3350, 3030, 3020, 1670, 1650, 1610, 1508, 820 cm^{-1}.

^{1}H NMR (300 MHz, $CDCl_3$):

1.65–1.78 m (H-1), 1.30 m (H-2), 1.75–1.98 m (H-3), 1.87–2.02 m (H-6), 1.26–1.58 m (H-7), 1.64 m (H-9), 1.42–1.88 m (H-11), 2.54 m (H-12), 1.56–1.91 m (H-13), 5.67 brs (H-15), 4.55 brs (H-17), 1.05 s (H-18), 7.43 brs (H-19), 3.57 brs (H-20), 6.85 d (8.0) (H-2'/6'), 7.12 d (8.0) (H-3'/5'), 2.99 brt (H-7'), 3.02 brt (H-8'), 2.40 s N-(CH_3)2-9'.

^{13}C NMR (75.03 MHz, $CDCl_3$):

27.9 (C-1), 30.1 (C-2), 30.9 (C-3), 44.9 (C-4), 72.6 (C-5), 30.9 (C-6),20.9 (C-7), 43.8 (C-8), 47.1 (C-9), 41.0 (C-10), 43.2 (C-11), 31.7 (C-12), 43.5 (C-13), 69.8 (C-14), 130.8 (C-15), 146.8 (C-16), 68.8 (C-17), 19.1 (C-18), 169.9, (C-19), 80.6 (C-20), 158.2 (C-1'), 114.9 (C-2'/6'), 129.2 (C-3'/5'), 125.0 (C-4'), 33.3 (C-7'), 52.5 (C-8'), 34.3 (N-(CH_3)2-9').

HRAPITOFMS *m/z* [M + H+] calculated for $C_{30}H_{40}N_2O_3$: 477.3117; found: 477.3122. Based on these data, the structure was concluded as in Figure.

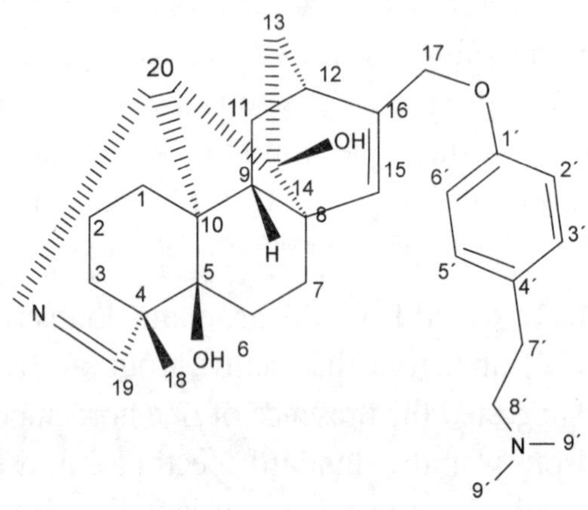

Hyphenated Techniques[28-32]

Chemical screening can involve simple TLC analysis but a much more selective and predictive method is HPLC coupled to different detectors (hyphenated techniques). This provides efficient separation of metabolites and valuable structure information at the same time, using only several micrograms of sample. Coupling HPLC with mass spectrometry (LC/MS) and UV diode array (LC/UV) detection has now been complemented by the connection of HPLC to nuclear magnetic resonance (LC/NMR). The latter allows the complete on-line structure determination of numerous plant metabolites.[a]

When compared to the classical use of UV, MS and NMR spectroscopy applied to pure natural products, ideally the integration of all these techniques in their hyphenated forms (LC/UV LC/MS and LC/NMR) in a single setup, with centralized acquisition of the spectroscopic data should permit the complete spectroscopic characterization of different metabolites in a mixture during a single analysis.

HPLC coupled with UV photodiode array detection (LC/UV), introduced by George and Maute (1982), allows the running of a chromatographic separation with simultaneous detection at different wavelengths. It has been used for more than two decades by phytochemists in the screening of the plant extracts and is now widely employed in many laboratories.

LC/MS is one of the most sensitive analytical methods and with the high power of mass separation of a MS detector, very good selectivities can be obtained. Moreover, this technique has the potential to yield information about the molecular weight as well as the structure of the analytes, using the MS/MS or MS^n capabilities of the analyser available. Because of the basic incompatibilities between HPLC and mass spectrometry, online linking of these instrumental techniques has been difficult to achieve; to cope with these problems, different LC/MS interfaces have been conceived. Each of these interfaces has its own characteristics and range of applications and several of them are suitable in phytochemistry. For the analysis of natural products, the most commonly used LC/MS interfaces are atmospheric pressure chemical ionization (APCI) and electrospray (ESI). Each of these interfaces has its own characteristics and range of application but their combined use permits the analysis of small nonpolar natural products to very large polar molecules. Direct on-line coupling of a NMR spectrometer as a detector for chromatographic separation has required the development of special interfaces called flow probes, as well the coupling of liquid chromatography (LC) or high-pressure liquid chromatography (HPLC) with such technologies as MS, MS/MS, and NMR has permitted the construction of devices that allow the injection of a crude sample, separation of the sample using automatically determined optimized conditions, and on the fly spectrometry or spectroscopy followed by CASE for each set of acquired spectra.

powerful method in the structure elucidation of natural products is to couple HPLC,MS and NMR. Pullen *et al.* were the first to construct a hyphenated method, which coupled HPLC-NMR-MS for analysis of a mixture. HPLC-NMR-MS in natural product research is a fast and powerful method that provides lots of structural information. HPLC-NMR-MS can be used at an early stage for identification of known compounds before bioassay-guided fractionation starts, so called dereplication. There are several operational conflicts to overcome when coupling HPLC with NMR and MS, such as solvent compatibility, instrumental sensitivity and magnetic field effects. However, the major obstacles for most research groups is probably the price of this hyphenated instrumentation, location and the expertise needed for operating.

REFERENCES

1. Mabry,T.J. (2001). "Selected topics from forty years of natural products research: Betalains to flavonoids, antiviral proteins, and neurotoxic nonprotein amino acids." *J. Nat. Products*, 64 (12), 1596–1604.
2. Nakanishi, K. (1999). An historical perspective of natural products chemistry. *Comprehensive Nat. Prod. Chem.* 8, xxi–xxxviii.
3. Nakanishi, K. (1999). An historical perspective of natural products chemistry. In. Ushio, S. (Ed.). *Comprehensive Natural Products Chemistry*, Vol. 1. Elsevier Science B.V., Amsterdam, pp. 23–40.
4. Yu, D., Chen,Y. and Liang, X. (2001). "Structural chemistry and biological activities of natural products from Chinese herbal medicines." *Emerging Drugs, 1* (Molecular Aspects of Asian Medicines). pp. 215–321.
5. Schrader, B. (1995). *Infrared and Raman Spectroscopy.* VCH, Weinheim.
6. Smith, E. and Dent, G. (2005). *Modern Raman Spectroscopy–A Practical Approach*, John Wiley & Sons Ltd, England.
7. Purcell, E.M., Torrey, H.C. and Pound, R.V. (1946). "Resonance absorption by nuclear magnetic moments in a solid." *Phys. Rev.* 69, 37–38.
8. Bloch, F.,Hansen, W.W. and Packard, M.E. (1946). "Nuclear induction." *Phys. Rev.* 69, 127.
9. Abragam, A.. (1961). *The Principles of Nuclear Magnetism.* Oxford University Press, London.
10. Ernst, R.R., Bodenhausen, G. and Wokaun, A. (1987). *Principles of Nuclear Magnetic Resonance in One and Two Dimensions.* Oxford Science Publications/Clarendon Press, Oxford.
11. Proctor, W.G. and Yu, F.C. (1950). "The dependence of a nuclear magnetic resonance frequency upon chemical compound." *Phys. Rev.* 77, 717.
12. Dickinson, W.C. (1950). "Dependence of the F19 nuclear resonance position on chemical compound". Phys. Rev. 77, 736.
13. [a] Kalinowski, H.O., Berger, S., and Braun, S. (1984). *^{13}C-NMR-Spektroskopie.* Georg Thieme verlag, stuggart.
14. Marshall, J. L. (1983). "Carbon-carbon and carbon-proton NMR couplings: Application to organic stereochemistry and conformational analysis." *MSA 2.* VCH, Deerfield Beach, Florida.

15. Neidig, K.P., Geyer, M., Gorler,A., Antz, C., Saffrich, R., Beneicke,W. and Kalbitzer, H. R. (1995). "Aurelia, a program for computer-aided analysis of multidimensional NMR spectra." *J. Biomol. NMR.* 6. 255–270.

16. Lindel, T., Junker, J. and Köck, M. (1997). "COCON: From NMR correlation data to molecular constitutions." J. Mol. Model. 3, 364-368.

17. Steinbeck, C. (2001). "The automation of natural product structure elucidation." *Curr. Opin. Drug Discov. Devel.* 4(3), 338–342.

18. Hemmer, M.C. and Gasteiger, J. (2000). "Prediction of three-dimensional molecular structures using information from infrared spectra." *Anal. Chim. Acta.* 420, 145–154.

19. Wolfender, J. L., Maillard, M., Marston, A. and Hostettmann, K., (1992). "Mass spectrometry of underivatised naturally occuring glycosides." *Phytochem. Anal.* 3, 193.

20. Takayama, M., Fukai, T., Noruma, T. and Yamauchi, T. (1991). "Formation and fragmentation of the $[M + Na]^+$ ion of glycosides in fast atom bombardment mass spectometry." *Org. Mass Spectrum.* 26, 655.

21. Grossert, J. S. (2001). "A retrospectro view of mass spectrometry and natural products—sixty years of progress, with a focus on contributions by R. Graham Corks." *Int. J. Mass Spectrum.* 212, 65.

22. Cuyckens, F. and Claeys, M. (2004). "Mass spectrometry in the structural analysis of flavonoids-I." *J. of Mass Spectrum.*, 39, 1.

23. Stobiecki, M. (2000). "Application of mass spectometry for identification and structural studies of flavonoid glucosides." *Phytochemistry.* 54, 237.

24. Siddiqui, T. N. Waheed, J. Lucke and Voelter, W. Z. (1975). "Mass spectrometry in the structural analysis of flavonoids-I." *Naturforsch.* 306, 961.

25. Siddiqui, T. N. Waheed, J. Lucke and Voelter, W. (1975). "Mass spectrometry in the structural analysis of flavonoids-II." *Chemiker-Ztg.* 99, 504.

26. Ikan, R., Crammer, B., (2002). "Organic chemistry: Compound detection." In: *Encyclopedia of Physical Sciences and Technology*, 3rd (edn). 11: 453–496.

27. Gao, L., Wei, X. and Yang, L. (2004). "A new diterpenoid alkaloid from a Tibetan medicinal herb *Aconitum naviculare* Stapf." *Journal of Chemical Research.* 307–308.

28. Wolfender, J-L., Rodriguez, S. & Hostettmann, K. (1998). "Liquid chromatography coupled to mass spectrometry and nuclear magnetic resonance spectroscopy for the screening of plant consituents." *Journal of Chromatography A.* 794. 299–316.

29. Hansen, S.H., Jensen, A.G., Cornett, C., Bjørnsdottir, I., Taylor, S., Wright, B. and Wilson, I. D. (1999). "High-performance liquid chromatography on-line coupled to

highfield NMR and mass spectrometry for structure elucidation of constituents of *Hypericum perforatum* L." *Analytical Chemistry*. 71, 5235–5241.

30. Lommen, A., Godejohann, M., Venema, D.P., Hollman, P.C.H. and Spraul, M. 2000. "Application of directly coupled HPLC-NMR-MS to the identification and conformation of quercetin glycosides and phloretin glycosides in apple peel." *Analytical Chemistry*. 72, 1793–1797.
31. Wilson, I.D. (2000). "Multiple hyphenation of liquid chromatography with nuclear magnetic resonance spectroscopy, mass spectrometry and beyond." *Journal of Chromatography A.* 892, 315–327.
32. Holt, R.M., Newman, M.J., Pullen, F.S., Richards, D.S. and Swanson, A.G. (1997). "High performance liquid chromatography/NMR spectroscopy/mass spectrometry: further advances in hyphenated technology." *Journal of Mass Spectrometry.* 32, 64–70.

4

INSIGHTS INTO NATURAL PRODUCTS

The aim of this chapter is to discuss some unique aspects of natural products. Since the understanding and comprehension of natural phenomena is limited, not everything is being appreciated—A quotable one being the fire flies. Why do they glow?

Layers of reflector cells are assembled around a layer of light-producing cells called photocytes, which are composed of tiny organs called peroxisomes. The peroxisomes contain three chemicals, viz., the protein luciferin (a protein), luciferase (an enzyme) and adenosine triphosphate (ATP). The mitochondria is stimulated with nitric oxide which fills the peroxisomes with oxygen. This causes luciferin and luciferase to glow until nitric oxide dissipates.

This chapter gives a sampling into the various facets of natural products and their functions.

CHEMICAL ECOLOGY

Chemical ecology is the study of the chemicals involved in the interactions of living organisms. It focuses on the production of and response to signalling molecules (i.e., semiochemicals) and toxins. Chemical communication is of particular importance among social insects including bees, wasps, and termites, as a means of communication essential to social organization. In addition, this area of ecology deals with studies involving defensive chemicals, which are utilized to deter potential predators, which may attack a wide variety of species.

Chemicals in the environment of an insect mediate much of its behaviour. By turning these chemicals to our own advantage, it is often possible to attract pest insects to traps or baits, or to repel them from our homes, our crops, or our domestic animals. Behavioural messages are delivered by a wide array of chemical compounds. As a group, these compounds are known as semiochemicals (*semeon* means a signal in Greek) and it is a generic term used for chemical substance or mixtures that carry a message. The semiochemicals are subdivided into two major groups, pheromones and allelochemicals, depending on whether the interactions are intraspecific or interspecific, respectively.

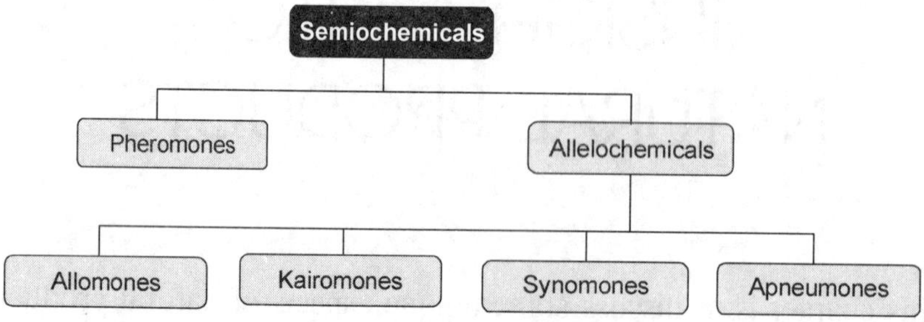

Figure 4.1 Classification of semiochemicals

The semiochemicals can be divided into several subgroups that distinguish different categories on the basis of their biological purposes (Figure 4.1). Substances that elicit a response within species are named pheromones. The word pheromone is based on the Greek word *pherein* (to transport) and *hormone* (to stimulate). The pheromone class is further split into several subclasses: sex, alarm, aggregation, trail and so forth (Figure 4.2).

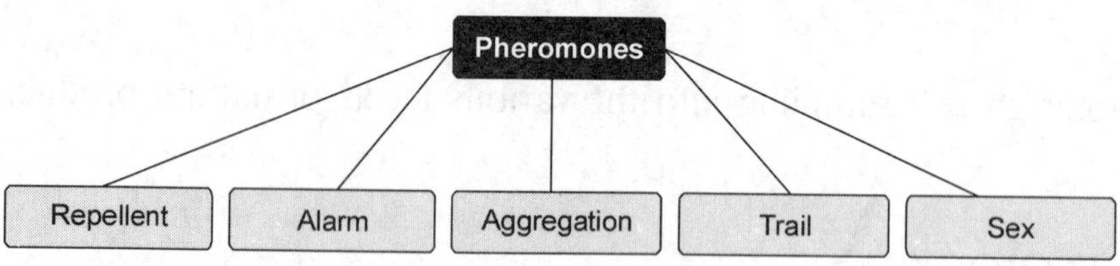

Figure 4.2 Classification of pheromones

One example from the repellent group is the male of the desert locust, *Schistocerca gregaria,* which in the gregarious phase releases phenylacetonitrile as a courtship inhibition pheromone (repellent) to ensure its reproductive success by preventing the female from mating with competing males.

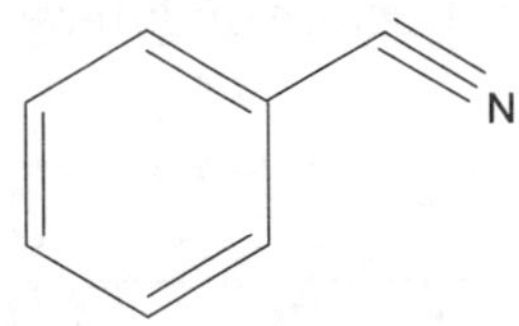

Phenylacetonitrile

When attacked by a predator, some species release a volatile substance that can trigger flight or aggression in members of the same species. In response to attack by natural enemies, most aphid species release an alarm pheromone that causes nearby conspecifics to cease feeding and disperse. The primary component of the alarm pheromone of aphid species studied is (*E*)-β-farnesene.[1]

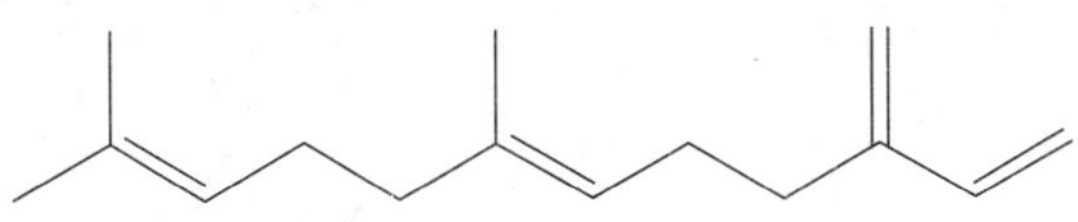

(*E*)-β-farnesene

Male-produced sex attractants have been called aggregation pheromones. They usually result in the arrival of both sexes at a calling site, increase in density of conspecifics surrounding the pheromone source. The banded alder borer, *Rosalia funebris,* is a long-horned beetle and the *(Z)*-3-decenyl *(E)*-2-hexenoate has been identified as a male-produced aggregation pheromone. [2]

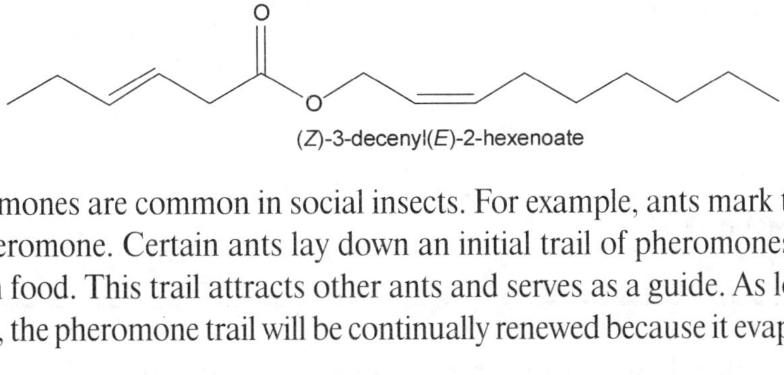

(*Z*)-3-decenyl(*E*)-2-hexenoate

Trail pheromones are common in social insects. For example, ants mark their paths with this type of pheromone. Certain ants lay down an initial trail of pheromones as they return to the nest with food. This trail attracts other ants and serves as a guide. As long as the food source remains, the pheromone trail will be continually renewed because it evaporates quickly.[3]

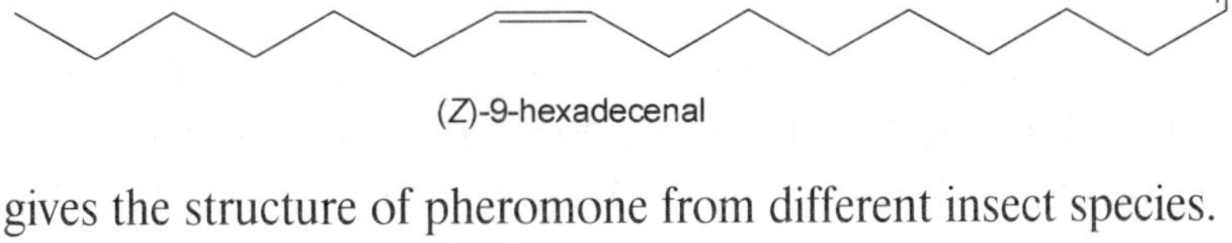

(*Z*)-9-hexadecenal

Table 4.1 gives the structure of pheromone from different insect species.

Table 4.1 Structure of different pheromones

Source (Insect)	Structure(s) of pheromone
1. Common housefly *(Musca domestica)*	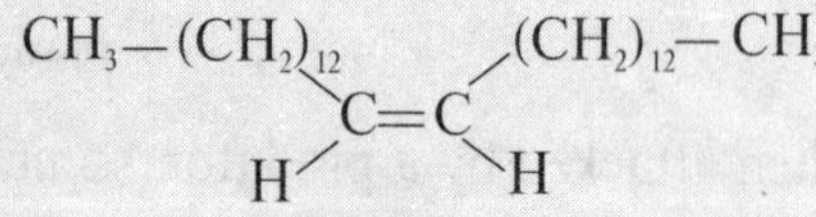
2. Honeybee *(Apis mellifera)*	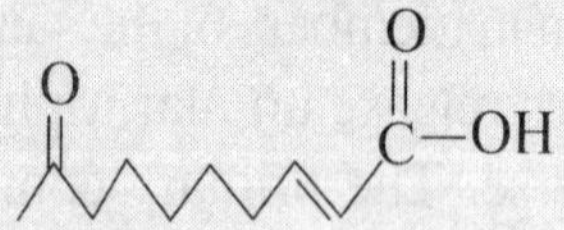
3. Gypsy moth *(Lymantria dispar)*	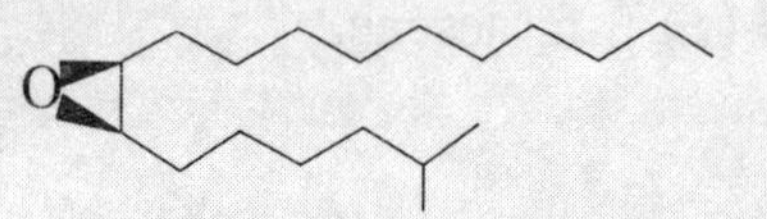
4. Japanese beetle *(Popillia japonica)*	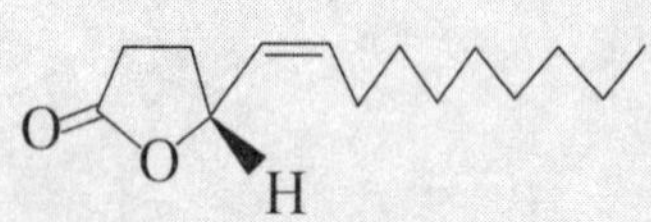
5. American cockroach *(Periplaneta americana)*	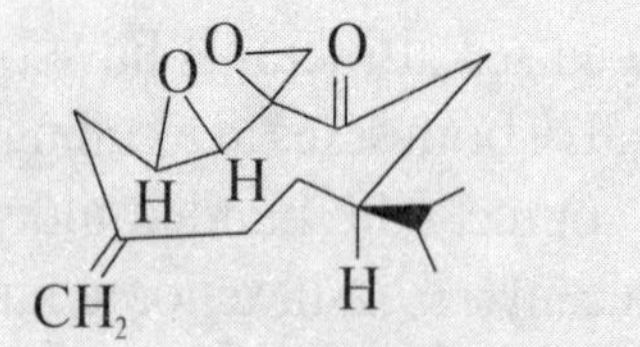
6. Spruce Budworm *(Choristroneura fumiferana)*	 96%  4%
7. Cabbage looper moth *(Trichoplusia ni)*	

A database : pherobase (www.pherobase.com) is a free database with over 30,000 entries, along with the analytical data available for a few. A snapshot of the database is given in Figure 4.3.

Chemicals, which transmit messages between different species, are called allelochemicals. These are also divided into different groups depending on their functions. Varied secondary metabolites are produced by plants with the sole aim of interactions with other living

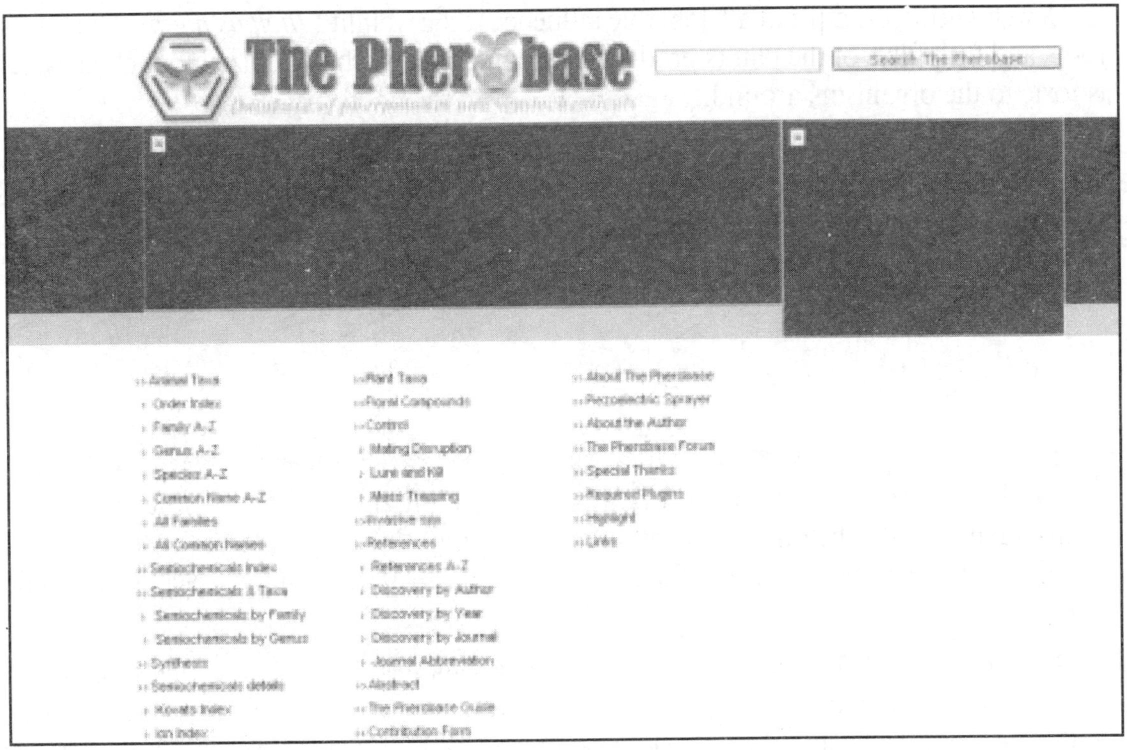

Figure 4.3 A snapshot of the pherobase web page

organisms in the environment. This might be for self-defence, sexual attraction, symbiosis, and development. The term allelopathy coined by Molisch in1937 was meant to define the direct/indirect effect of transfer of chemicals from plants, for which the scope has further widened to include algae, fungi, and the various microorganisms as well as higher plants. The term "Allelopathy" encompasses both inhibitory and stimulatory activity. Allelochemicals can be spread into the environment by any of the following ways, exudation from roots, volatilization/leaching from leaves or plant litter on ground by precipitation.

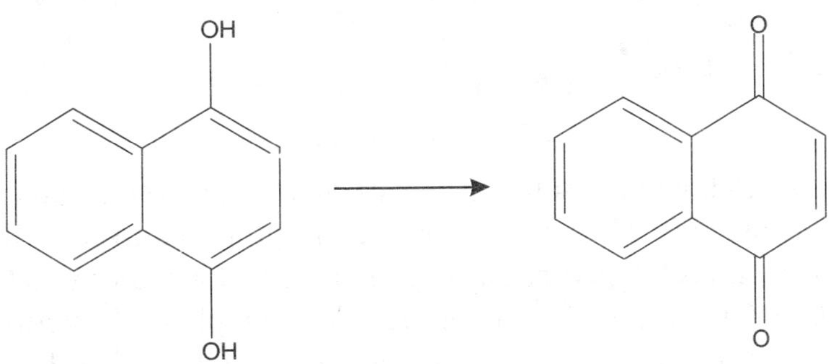

Hydroquinone Juglone

A well-known example of allelopathic influence is the walnut (*Juglans nigra*). The leaves, roots and fruit hulls of the plants produce hydroquinone which is oxidized to juglone which is toxic to the organisms around.

Coumarin and its derivative scopoletin have been reported to inhibit seed germination and growth of plants[4] and coumarin blocked mitosis in *Allium cepa* (onion).[5] Psoralen inhibits lettuce seed germination at a concentration of 1 ppb.[6]

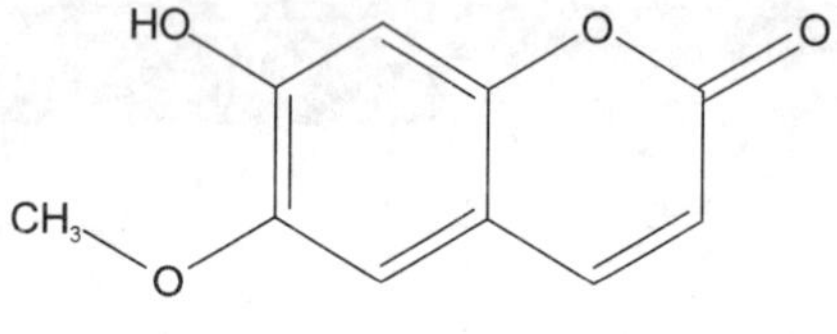

Scopoletin

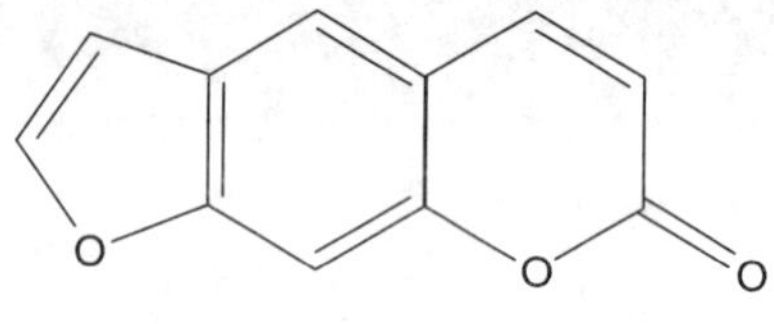

Psoralen

Antibiotics produced by microorganisms to inhibit the growth of other species of microorganisms also belong to this allomones category. Judging from the vast number of antibiotics that have been characterized, chemical warfare at the bacterial level must be quite extensive.

Four types of allelochemics are recognized: allomones, kairomones, synomones, and apneumones. An **allomone** is a chemical produced or acquired by an organism that evokes in the receiver a behavioural or physiological response that is adaptively favourable to the emitter but not to the receiver.[7, 8] Feeding inhibitors such as azadirachtin (a tetranortriterpenoid isolated from the seeds of neem tree) are examples.

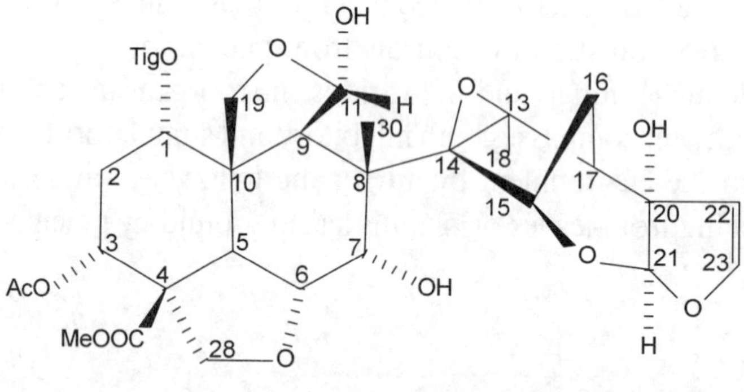

Azadirachtin

In contrast to the allomones, some chemicals benefit the receiver rather than the emitter in interspecific interactions. The term **kairomone** was proposed to cover the chemicals that mediate these interactions.[8] Bark beetle aggregation pheromones are exploited as kairomones by several predators as means of locating prey. An example is *Anopheles gambiae sensu stricto*, a malaria-causing mosquito, that uses human volatiles such as lactic acid, ammonia and several carboxylic acids for host orientation.

Nordlund and Lewis[7] proposed the third term **synomone** for chemicals that mediate mutualistic interactions, i.e., they affect both the emitter and the receiver positively. They also proposed the fourth term **apneumone** for chemicals that are emitted by a non-living material and evoke a behavioural or physiological reaction that is adaptively favourable to a receiving organism but detrimental to an organism of another species that may be found in or on the non-living material. There are also examples of multifunctional semiochemicals.

As we identify the specific plant components involved and their actions, it will be easier to incorporate into the plant, the capacity to produce the desired chemical. There is a further possibility of identifying additional natural chemicals that may be useful as pest control materials, through the use of natural products or products of industrial synthesis patterned after the natural products. It is clear that allelochemicals are involved in these complex processes and they hold promise for even greater applications. Expanded research in the new biotechnologies offers great potential for further development of allelochemicals. Interactions among plants and other organisms have long been recognized and described. For example, gardeners long ago observed that tomatoes grow/yield poorly under black walnut trees. Sorghum plant residues inhibit the growth of many weeds. Pyrcthrum and neem plants have definite insecticidal properties. Certain varieties of crop plants exhibit resistance to insects and diseases. We now know that many such relationships involve chemicals, known as allelochemicals, produced by the plants or other organisms. On a chemical and molecular basis, we are now beginning to unravel why these interactions among various organisms occur.

In the Gramineae, the cyclic hydroxamic acids 2,4-dihydroxy-1,4-benzoxazin-3-one (DIBOA) and 2,4-dihydroxy 7-methoxy 1,4-benzoxazin-3-one (DIMBOA) take part in the defence against insects and microbial pathogens. [9,10]

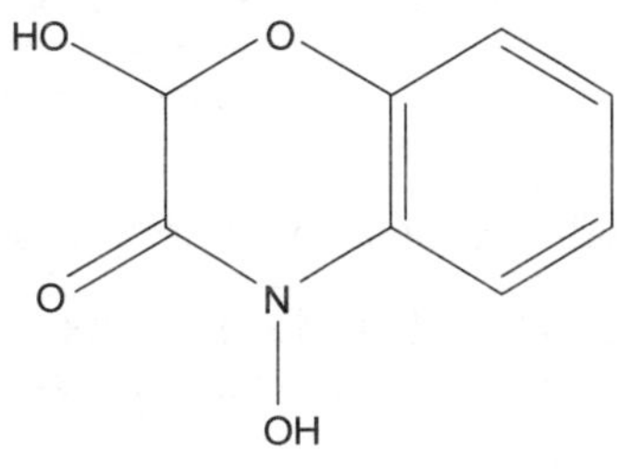

DIBOA

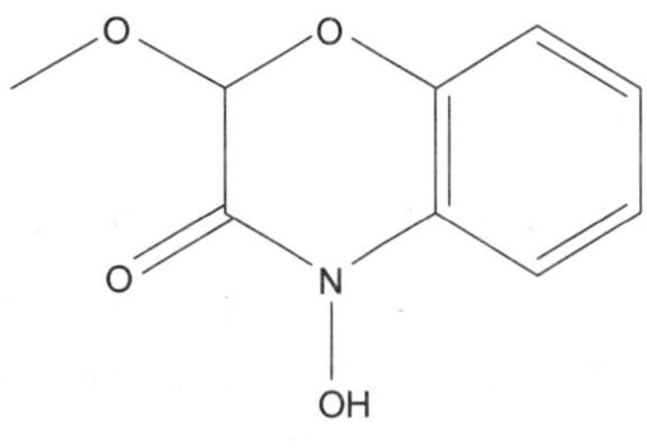

DIMBOA

Antibiotics, antineoplastics, herbicides, and insecticides often originate from plant and microbial chemical defence mechanisms.[11] Secondary metabolites, once considered unimportant products, are now thought to mediate chemical defence mechanisms by providing chemical barriers against animal and microbial predators.[12] Plants produce numerous chemicals for defence and communication, and can elicit their own form of offensive chemical warfare by targeting the proliferation of pathogens. These chemicals may have general or specific activity against key target sites in bacteria, fungi, and viruses. *p*-hydroxybenzaldehyde which is present in the epicuticular layer of the *Sorghum* cultivar 65-D was found to cause 90% inhibition of locust feeding.

Interactions between insect pheromones and semiochemicals (molecules that carry signals from one organism to another) from the host plant have been well-known. Some insects acquire host plant chemicals to use them as sex pheromones or sex pheromone precursors. Other host plant volatiles can induce the production or release of pheromones in certain insects and often synergize or enhance insect responses to sex pheromones. Host compounds can also have an inhibitory or repellent effect, interrupting the response of insects to their own pheromone. In other cases, host-derived compounds resulting from herbivorous attack can attract predators to the attacking insect and therefore serve as a defence mechanism for the plant.

One notable example being the monocrotaline—an alkaloid generated in the plant *Crotalaria spectabilis*—is converted into *(R)*-hydroxydanaidal a sex attractant pheromone of *Utetheisa ornatrix* (Arctiidae) which is depicted in Figure 4.4.

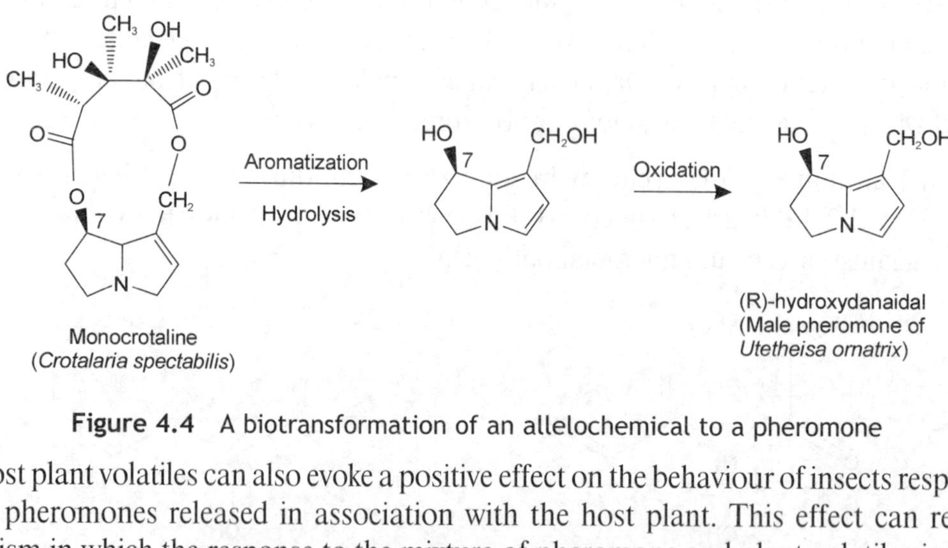

Figure 4.4 A biotransformation of an allelochemical to a pheromone

Host plant volatiles can also evoke a positive effect on the behaviour of insects responding to sex pheromones released in association with the host plant. This effect can result in synergism in which the response to the mixture of pheromone and plant volatiles is greater than the combined responses to the individual components. Synergism between plant semiochemicals and pheromones can contribute to more successful mate finding and therefore it is likely to play an important role in reproductive isolation. Host-odour enhancement of attraction responses to pheromones occurs in several insect orders.

From Table 4.2 we can understand the synergism between allelochemicals and aggregation pheromones, from Table 4.3 that between the plant volatiles and sex pheromones.

Table 4.2 Synergism between allelochemicals and aggregation pheromone

Host	Insect	Plant volatiles	Sex pheromone
Cotton	*Anthonomus grandis*	*trans*-2-hexenol, *cis*-3-hexenol, *n*-hexanol	Grandisol, grandisal
Palm, sugar cane (*Saccharum officinarum*), pineapple, banana	*Metamasius hemipterus sericeus*	Ethyl esters	5-methyl-4-nonanol, 2-methyl-4-heptanol
Apple, orange, stored grains, spices, seeds	*Carpophilus hemipterus*	Propanoic acid, butanoic acid, methanol, 2-propanol, *n*-heptanol, methyl butanoate, propanol	(2E,4E,6E,8E)-3,5,7-trimethyl-2,4,6,8-decatetraene
Cereal grain, flour	*Sitophilus oryzae*	Valeraldehyde, maltol, vanillin	Sitophinone
Palm	*Rhynchophorus phoenicis, Rhynchophorus vulneratus*	e.g., Alcohols and esters	3-methyl-4-octanol, 4-methyl-5-nonanol (rynchophorol)
Palm	*Rhynchophorus cruentatus*	Ethyl acetate	5-methyl-4-octanol (cruentol)
Wheat	*Carpophilus mutilatus*	Fermenting whole wheat bread dough	(3E,5E,7E)-5-ethyl-7-methyl-3,5,7-undecatriene
Coconut, palm, banana	*Rhynchophorus palmarum*	Ethyl acetate	(2E)-2-methyl-5-hepten-4-ol
Wheat	*Carpophilus obsoletus*	Propyl acetate	(2E,4E,6E,8E)-3,5,7-trimethyl-2,4,6,8-undecatetraene
Palm	*Rhynchophorus phoenicis*	Palm tissue volatiles	Rhynchophorol

(Contd.)

Table 4.2 (Continued)

Host	Insect	Plant volatiles	Sex pheromone
Palm	*Rhynchophorus ferrugineus*	Host plant volatiles	Ferrugineol (4-methyl-5-nonanol)
Oil palm (*Elaeis oleifera*)	*Oryctes rhinoceros*	Oil palm, fruit bunches	Ethyl-4-methyloctanoate
Coconut, palm, banana	*Rhynchophorus palmarum, Dynamis borassi*	Ethanol, ethyl acetate	Rhynchophorol
Palm, sugar cane	*Rhabdoscelus obscurus*	Ethyl acetate	2-methyl-4-octanol (2E)-6-methyl-2-hepten-4-ol, 2-methyl-4-heptanol
Fermenting aspen (*Populus tremula*) bark	*Drosophila borealis, Drosophila littoralis*	Host odours	Ethyl tiglate

Table 4.3 Synergism between allelochemicals and sex pheromone

Host	Insect	Plant volatiles	Sex pheromone
Zea mays	*Helicoverpa zea*	(Z)-3-hexenyl acetate	(Z)-11-hexadecenal, (Z)-11-hexadecenol, (Z)-9-hexadecenal, (Z)-7-hexadecenal, hexadecanal
Zea mays	*Cydia pomonella*	(Z)-3-hexenyl acetate	(E,E)-8, 10-dodecadienol (codlemone)
Beta vulgaris	*Spodoptera exigua*	Linalool, myrcene benzaldehyde	(Z,E)-9,12-tetradecadienyl acetate, (Z)-9-tetradecenol
Prunus padus	*Rhopalosiphum padi*	Benzaldehyde	Nepetalactol

(Contd.)

Table 4.3 (Continued)

Host	Insect	Plant volatiles	Sex pheromone
Japanese cedar (*Cryptomeria japonica*), Japanese cypress (*Chamaecyparis obtuse*)	*Anaglyptus subfasciatus*	Methylphenyl acetate	(R)-3-hydroxy-2-hexanone, (R)-3-hydroxy-2-octanone
Brassica oleracea subsp. *capitata*	*Plutella xylostella*	(Z)-3-hexenyl acetate, (E)-2-hexenal, (Z)-3-hexenol	(Z)-11-Hexadecenal, (Z)-11-hexadecenyl acetate, (Z)-11-hexadecenol
Pine	*Pissodes nemorensis*	Pine bolt odours	Grandisol, grandisal
Papaya	*Toxotrypana curvicauda*	Host fruit odours	2-methyl-6-vinylpyrazine
Cotton, tobacco, tomato	*Heliothis virescens*	(Z)-3-Hexenyl acetate	(Z)-11-Hexadecenal, (Z)-9-tetradecenal, hexadecanal, tetradecanal

CIRCADIAN SYSTEMS

Circadian systems are widespread endogenous mechanisms that allow organisms to time their physiological changes to predictable day/night cycles. They have evolved in a wide range of organisms, from cyanobacteria to mammals, indicating their importance in life processes. Among an enormous variety of 24 h rhythms that are controlled by the circadian system are nitrogen-fixation in cyanobacteria, olfactory responses in Drosophila and sleep patterns in humans. The rotation of the earth causes predictable changes in light and temperature in our natural environment. Accordingly, natural selection has favoured the evolution of circadian (from the Latin *circa*, meaning "about" and *dies*, meaning "day") clocks or biological clocks — endogenous cellular mechanisms for keeping track of time. These clocks impart a survival advantage by enabling an organism to anticipate daily environmental changes and thus tailor its behaviour and physiology to the appropriate time of the day. The clock is synchronized by the day–night cycle, allowing the organism to accommodate not only the daily cycles of light and dark attributable to the earth's rotation, but also the alteration in relative span of day and night caused by the tilting of the earth's axis relative to the sun. Thus, a circadian timing mechanism influences the biochemical processes in living beings.

People from indigenous cultures who use plants for medicines often have an intimate connection with the environment around them, including knowledge of seasonal patterns of plant potency. Further, some cultures take special care to plant, collect, or harvest at only certain times of the day, or during specific moon phases. Could harvest and collection practices and the influences of the moon phase affect the chemical composition and efficacy of plant-based medicines? The circadian clock regulates diverse aspects of plant growth and development and promotes plant fitness.

Plant rhythms have fascinated scientists for over a century, not surprisingly due to the myriad legends and folklore about the moon from indigenous cultures going back through millennia. The earliest scientific studies attempted to understand the "sleeping" behaviour of some plants whose leaves would fold flat against the stem at night and open during the day. All living organisms possess physiological processes and behaviours that follow rhythms involving a period of about 24 hours.[13] In humans and other mammals, an internal timing mechanism (often called the biological clock, or central oscillator), imposes circadian rhythms on physiological, biochemical, and molecular events. Plants, however, are unique, and their timing control mechanism is external, specifically the environment, i.e., the sum of all of the forces constantly acting upon the plant.

Circadian rhythms are the subset of biological rhythms with period, defined as the time to complete one cycle of ~24 h. This defining characteristic inspired Franz Halberg in 1959 to coin the term circadian, from the Latin words "circa" (about) and "dies" (day). A second defining attribute of circadian rhythms is that they are endogenously generated and self-sustaining, so they persist under constant environmental conditions, typically constant light (or dark) and constant temperature. Under these controlled conditions, the organism is deprived of external time cues, and the free-running period of ~24 h is observed. A third characteristic of all circadian rhythms is temperature compensation; the period remains relatively constant over a range of ambient temperatures. This is thought to be one facet of a general mechanism that buffers the clock against changes in cellular metabolism.

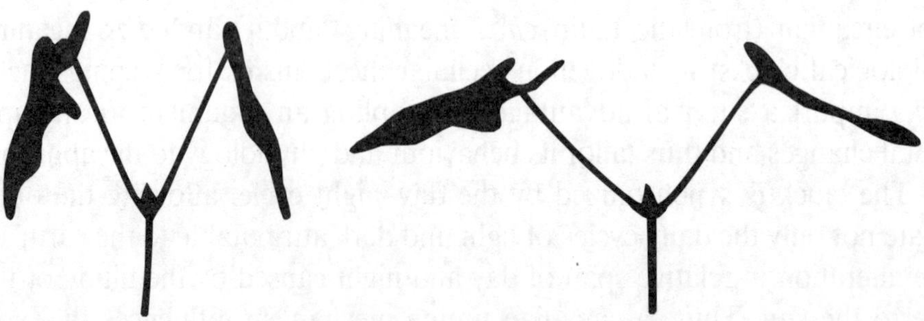

Figure 4.5 A circadian phenomena

Sleep movements of the plant *Phaseolus coccineus* corresponds to the presence and absence of light. The position of the primary leaves of a seedling at night is at the left and during the day is at the right (Figure 4.5). Greater success has been achieved in connecting auxin to circadian regulation of growth. The levels of free indole-3-acetic acid (IAA) and its conjugated form, IAA–aspartate, were shown to cycle in continuous light both in floral stems and in rosette leaves of *Arabidopsis*.[14]

There is some circumstantial evidence that other hormones may be involved in regulating circadian output. In carrot (*Daucus carota*), the levels of cytokinins (CKs) are under circadian control, [15] while in tobacco, CKs, as well as IAA and abscisic acid are rhythmic under diurnal condition. [16] It has been suggested that CKs may also be a part of an input pathway to the *Arabidopsis* clock. [17, 18] For human beings the "Jet Lag" that is associated with long-time flight travel (mostly to countries with large time interval) is due to the Circadian rhythm being affected.

METABOLITE–INSECT INTERACTION

Salix species are known to contain salicylates that influence herbivory, but the leaves also contain an array of other compounds that could modulate the feeding and fitness of insects that feed on the foliage. [19]

For example, *Salix pentandra L.* contains acetylated salicylates as well as chlorogenic acid and flavonoids and is a favoured host for Salix-specific insects but not polyphagous feeders, whereas another species, *S. phylicifolia* L., contains flavonoids but not salicylates, and leaves of this species are eaten by polyphagous species. Recent research on the European silver birch (*Betula pendula* Roth)[20] showed that increased production of antioxidant phenolics such as quercetin (2) and chlorogenic acid increased resistance of trees to ozone damage by scavenging free oxygen, hydroxyl radicals and hydrogen peroxide. Similar changes in the concentration of phenolics have been recorded in other species of trees.

For example, in needles of *Pinus nigra* Ait. increased production of phenolics is associated with an activation of the senescence process.[21]

These plants show a shift in the allocation of carbon for the production of chemical defence rather than growth. An increase in phenolics could make the needles more resistant to attack by insects. The activity of these phenolics could be very insect-specific. About 40 phenolic compounds have been identified in birch trees and gallotannin occurs in high concentrations in young leaves. The variation in the levels of the phenolic compounds has been shown to modulate the growth of the polyphagous larvae *Epirrita autumnata* Bkh.[22]

One of the phenolics, 1-*o*-galloyl-*D*-glucopyranose (glucogallin) was investigated to see if it was detrimental to larvae and was shown not to be an effective defence compound.[23]

Results of some of the earlier research on the effects of quercetin and quercetin glycosides on insects were collated by Harborne and Grayer[24] and Simmonds.[25] The responses of insects to these compounds vary greatly. For example, rutin is a phagostimulant to *Heliothis virescens* [26] and the locusts *Schistocerca americana*,[27] *Schistocerca albolineata* (Thomas) and *Melanoplus differentialis* (Thomas).[28]

Its influence on larvae of *Helicoverpa zea*, *H. armigera*, *Spodoptera littoralis*, *S. exiqua* and *S. exempta* depends on the concentration tested; at concentrations between 10^{-4} M and 10^{-5} M it stimulated feeding but at higher concentration it was a deterrent . Thus the role of quercetin derivatives in insect–plant interactions appears to be complex. This complexity is illustrated by the research on the host selection behaviour of Yponomenta larvae that consume plants containing rutin.[29]

Caterpillars use neurons in taste sensilla on their mouthparts to detect compounds in food, many of these compounds stimulate neurons that elicit a behavioural response resulting in either food acceptance or rejection.[30]

However, rutin might not be used by the insects to select food since electrophysiological studies on the neurons in the maxillary taste sensilla of larvae of Yponomenta showed that none of the neurons in either the lateral or medial maxillary styloconic sensilla responded to stimulation with rutin.

The "signals" that are released by a plant under stress can influence the biosynthesis of flavonoids and thus alter the profile of compounds in that plant. For example, when aphids feed on sorghum leaves, the leaves develop red spots. These red spots are associated with the expression of anthocyanins, a response that is also associated with water-stress. The induction of the accumulation of the anthocyanin red spots is thought to be a consequence of the de-activation of phytoalexins (3-deoxy-anthocyanidins) that occurs after insect or fungal attack, the elicitors being oligosaccharides that are produced by enzymes involved in the degrading of plant cell walls. The resistance of plants to pathogens has been attributed to an increase in anthocyanins.[31]

SYNTHESIS OF NATURAL PRODUCTS

From a purely chemical perspective, the total synthesis of natural products presents a challenge proving ground for synthetic methodology. One of the chief complaints about natural products in drug discovery is the issue of availability. The sustainable production of sufficient drug to supply demand in an economical fashion is a critical factor in the viability of any pharmaceutical candidate. The construction of complex organic molecules from simple chemicals is a multifaceted endeavour. Initially, a strategy must be devised that sequences chemical transformations, known or projected novel ones, in a way that could plausibly lead to the target structure. The number of steps and the yield of each transformation determine the efficiency of a successful synthesis. The structure and properties of the target molecule define the extent of the challenge. Although several metrics have been advanced to quantify the complexity of synthesis targets and, thus, the degree of difficulty involved, none is universally appropriate. Features of target structures that increase the challenge of a chemical synthesis include the following: i) instability, either thermodynamic instability or lability in the presence of common reagents or solvents; ii) the number of functional groups, particularly diverse ones displaying disparate reactivity; iii) the density of these functional groups; iv) the incidence of stereocentres; v) the number and types of rings; and often vi) the overall size of the molecule.

Many drugs derived from microorganisms can be successfully generated by fermentation, e.g., daptomycin.[32] However, total synthesis also plays an important role in alleviating the problem of supply. Galantamine (Alzheimer's) and mycophenolate sodium (immunosuppression), two natural product drugs which have recently reached the market, are both produced by total synthesis.[33] Semi-synthesis is another approach to alleviating problems of supply, with paclitaxel or taxol being a good example.

This clinically important anti-cancer drug is a natural product originally isolated from the bark of the Pacific yew tree (*Taxus brevifolia*) and its potential was recognized during research in the 1960s and 70s.[34] However, only 1 g of the metabolite could be extracted from 24 kg of dried bark. Since a 125-year-old tree was only capable of producing around 2 kg of bark and this species was relatively rare, the issue of supply was particularly acute in the case of this natural product and slowed its development as a drug. The total synthesis of taxol was completed by Holton and Nicolaou independently.[35, 36, 37] However, it was taxol's semi-synthesis from a natural congener, 10-deacetyl baccatin, more sustainably obtained from the needles of the *Taxus baccata*, which allowed for its large-scale manufacture and commercial supply.

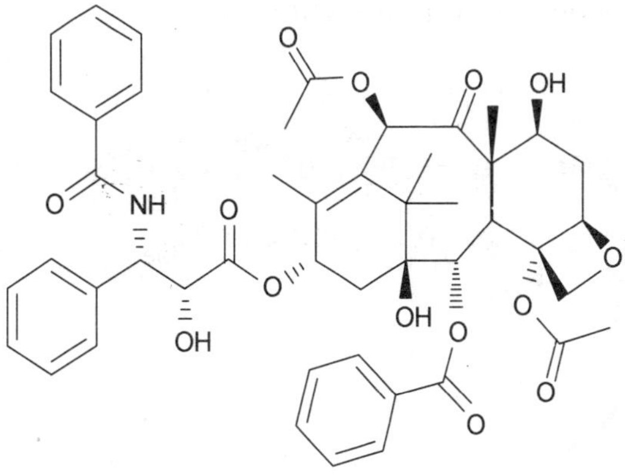

Taxol

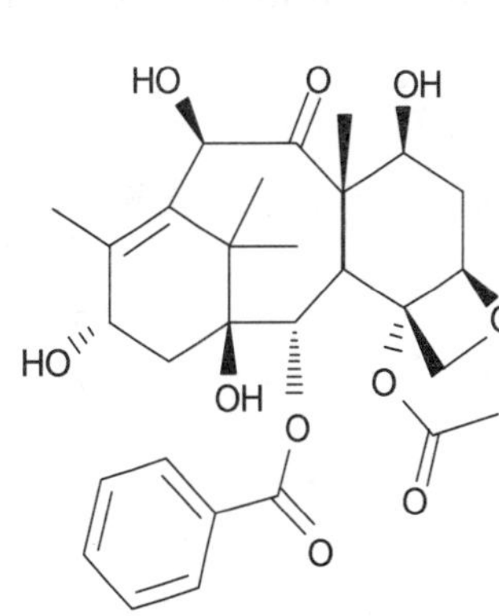

10-deacetyl baccatin

However, many organisms produce a vast array of secondary metabolites, many or all of which have no known role. This may be explained simply by accepting that whilst the role is not obvious to the outside observer, a role for a given metabolite may still exist, or at least have existed in the past. This view is being countered by a fascinating recent hypothesis put forward by Firn and Jones, the Screening Hypothesis.[38] This theory is based on the premise that potent biological activity is a rare property for any one molecule to possess. The authors therefore suggest that instead of carefully evolving a complex biosynthetic pathway to discreet potent metabolites, organisms endeavour to create natural product libraries, even if many of the metabolites eventually never play any role in the fitness of the organism. In this way the producer has an array of "chemical weapons" which will be screened during interactions with other organisms, one or more of which may confer an advantage to the producer. This theory consciously draws parallels with the high-throughput screening processes employed by man in the search for biologically useful compounds. The economics of natural selection requires that the production of secondary metabolites must not 'cost' the organism more than the benefit derived. In most of the cases, the role of libraries of secondary metabolites produced by organisms, has not been explored. The promiscuity of enzymes involved in secondary metabolite production is becoming increasingly accepted. The screening hypothesis remains contentious, however.[39,40]

TOTAL SYNTHESIS OF A MACROLIDE

The pamamycins are a class of 16-membered macrolide homologues that have been isolated from various *Streptomyces* species.[41,42] Next to displaying pronounced autoregulatory, anionophoric, and antifungal activities, several members of this family have been shown to be highly active against gram-positive bacteria including multiple antibiotic-resistant strains of *Mycobacterium tuberculosis.*[43] The retrosynthetic pamamycin plan is shown in Figure 4.6. The total synthesis of this macrolide is shown in Figure 4.7.

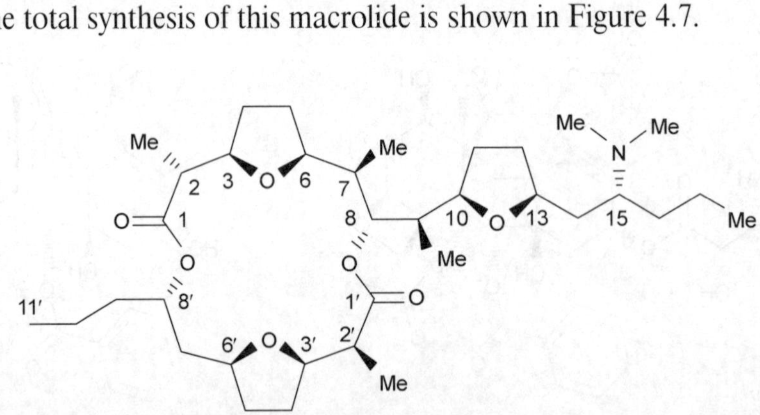

Pamamycin 607

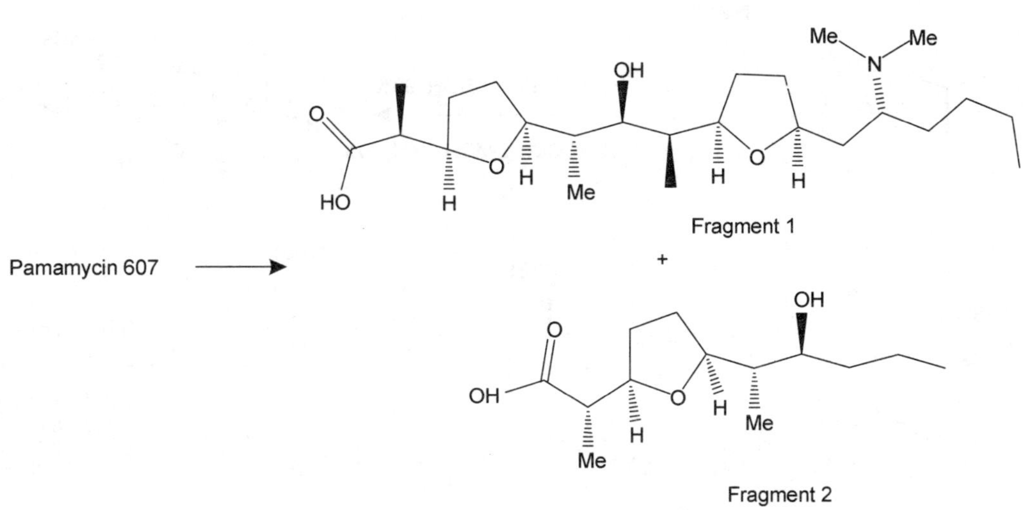

Figure 4.6 Retrosynthetic plan of pamamycin 607[44]

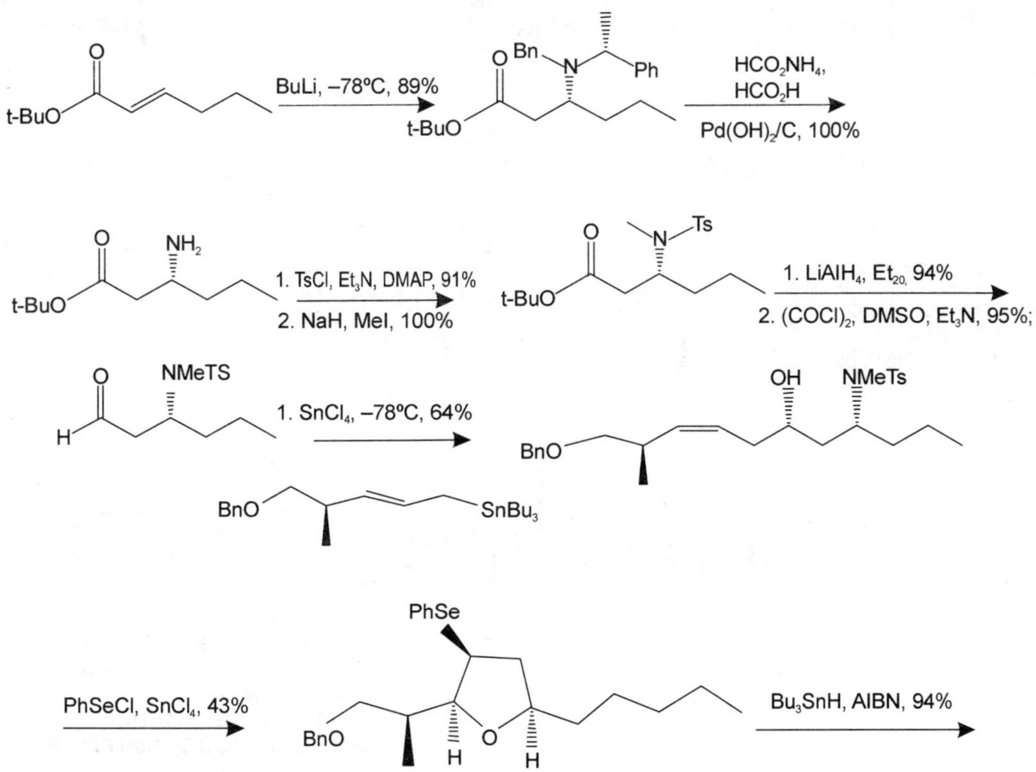

Figure 4.7 Total synthesis of the antibiotic macrolide Pamamycin 607 *(Continues)*

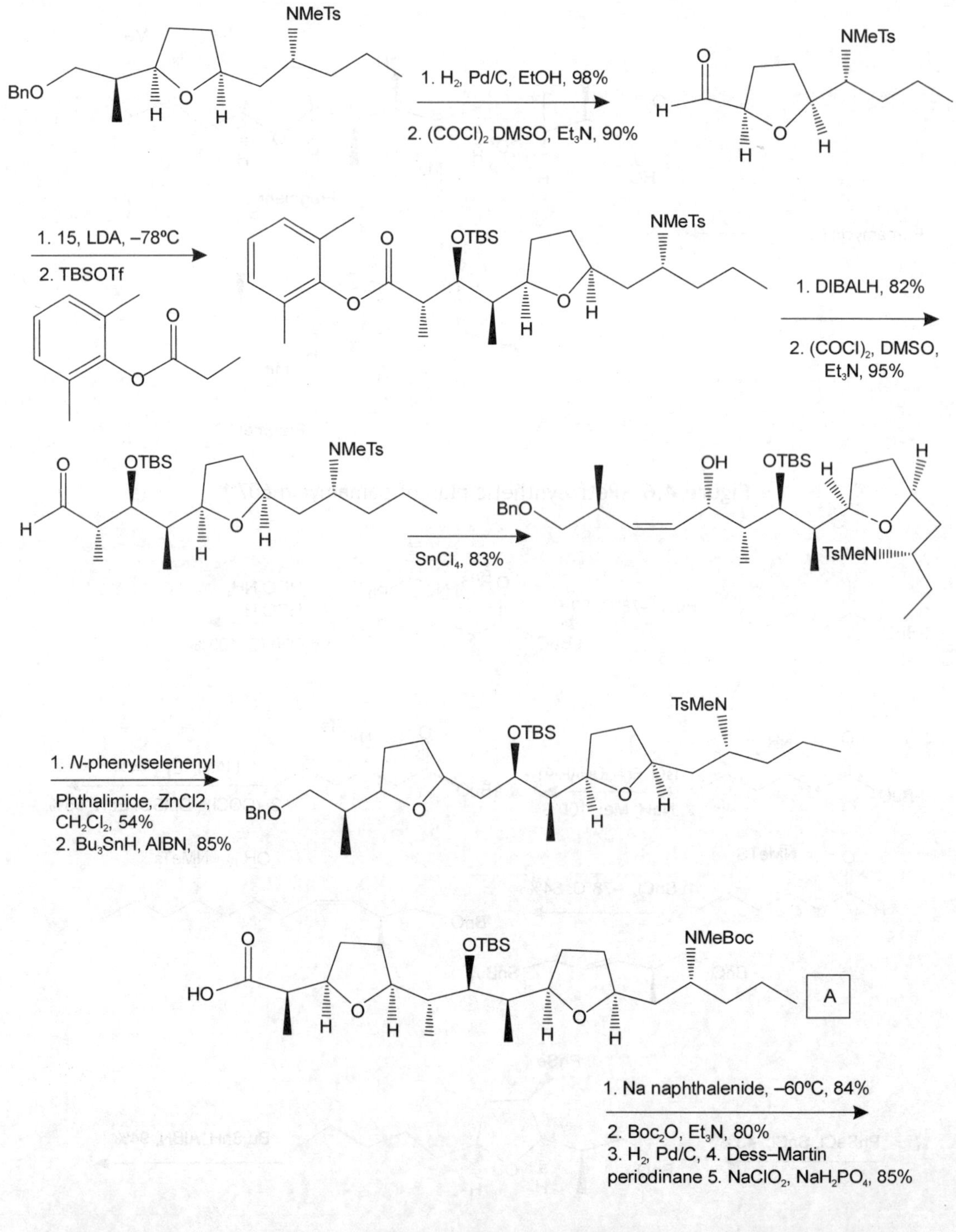

Figure 4.7 Total synthesis of the antibiotic macrolide Pamamycin 607 *(Continues)*

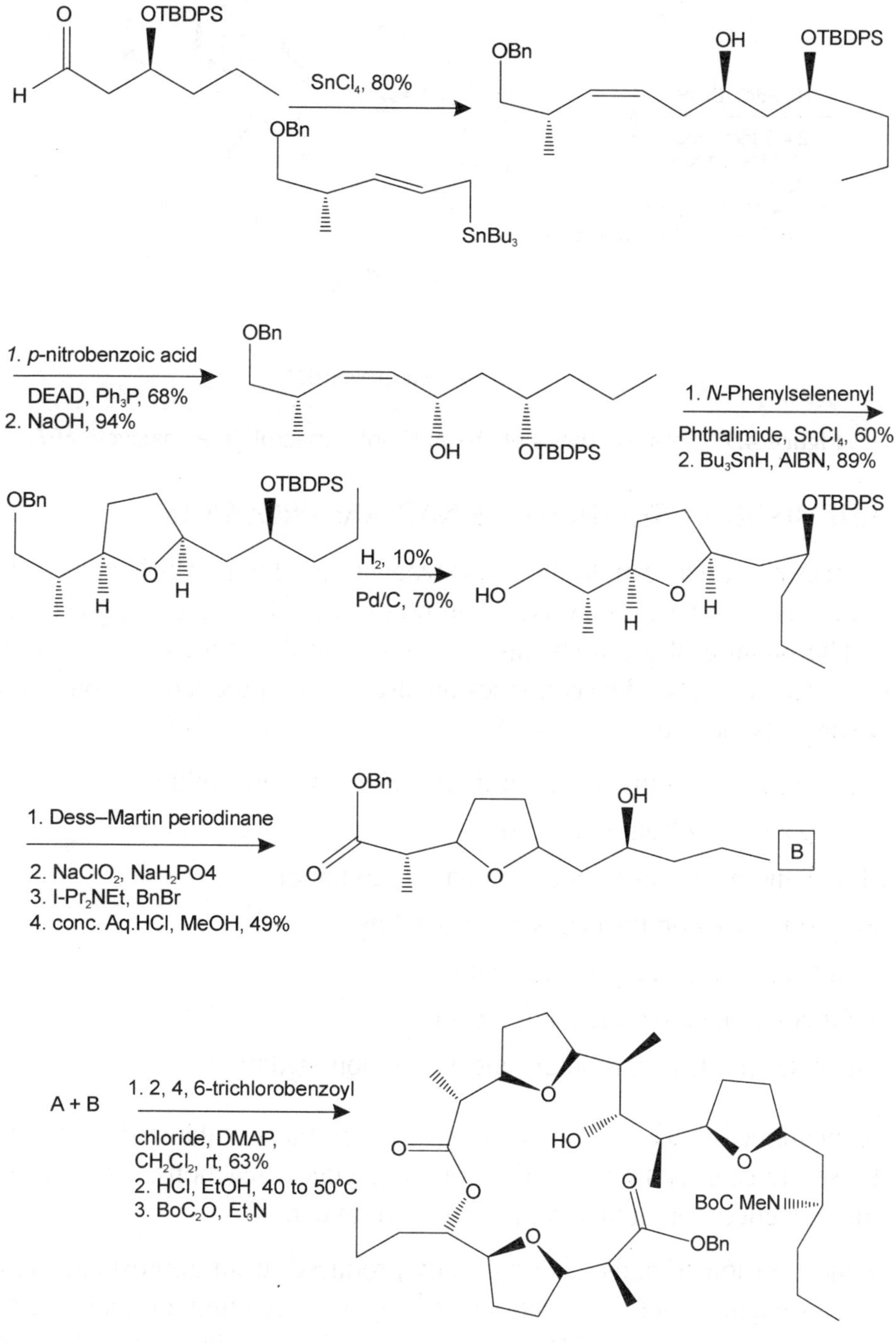

Figure 4.7 Total synthesis of the antibiotic macrolide Pamamycin 607 *(Continues)*

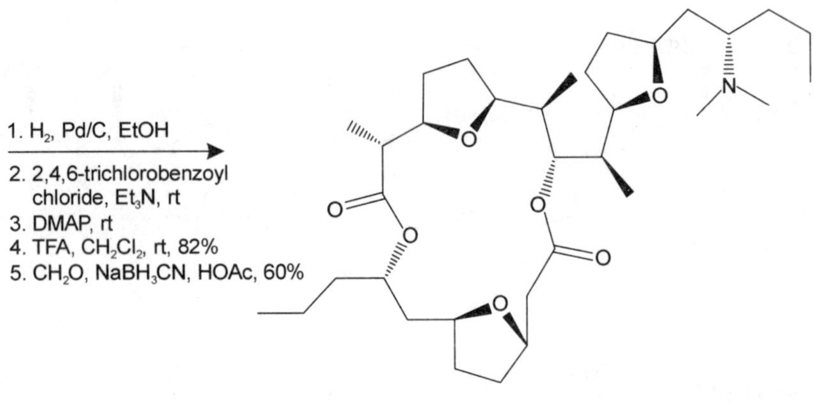

Figure 4.7 Total synthesis of the antibiotic macrolide Pamamycin 607

GREEN CHEMISTRY IN SYNTHESIS OF NATURAL PRODUCTS

Though synthesizing organic molecules using reagents/conditions which might be hazardous to the environment has been widely used till now, the focus has shifted to green chemistry in synthesis. The premise of green chemistry is based on the reduction in waste materials, hazardous impact and also the economics involved in the production. Some of the green chemistry supports include:

1. Use of microbial transformation/tissue/cell suspension culture
2. Using water as solvent in reactions
3. Ultrasound/microwave/photo-irradiation conditions
4. Solid-state reaction through simple grinding
5. Use of recyclable catalyst like Zeolite
6. Multi-component or single-pot reaction
7. Use of supercritical carbon dioxide as reaction medium

All the mentioned-processes can be followed if it results in enhanced yields and reaction rates.[45] These are ecofriendly unlike the normal chemical reactions, wherein most of the reagents have ill effects on human beings or environment.

The major portion of malic acid currently produced at an approximate 10,000 t/y is racemic, because it originates from petrochemically produced fumaric acid. The L-form can also be generated from fumaric acid by its hydration with immobilized cells of *Brevibacterium* or *Corynebacterium*.

Chalcones are phytochemicals with antibacterial, antifungal, antitumour and anti-inflammatory properties. Most of the chalcones can be obtained in a matter of minutes[46] by mixing the corresponding benzaldehyde and acetophenone in the presence of solid NaOH in a mortar with pestle. The yields of crude product were in the range 81–94%(Figure 4.8).

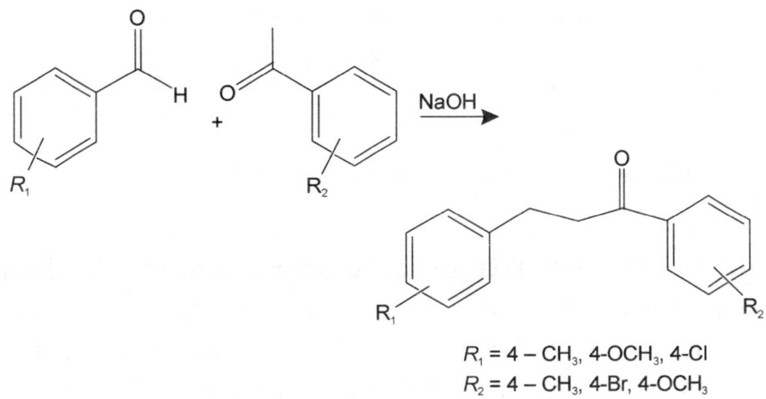

Figure 4.8 Solid state reaction of aldehyde and acetophenone derivatives

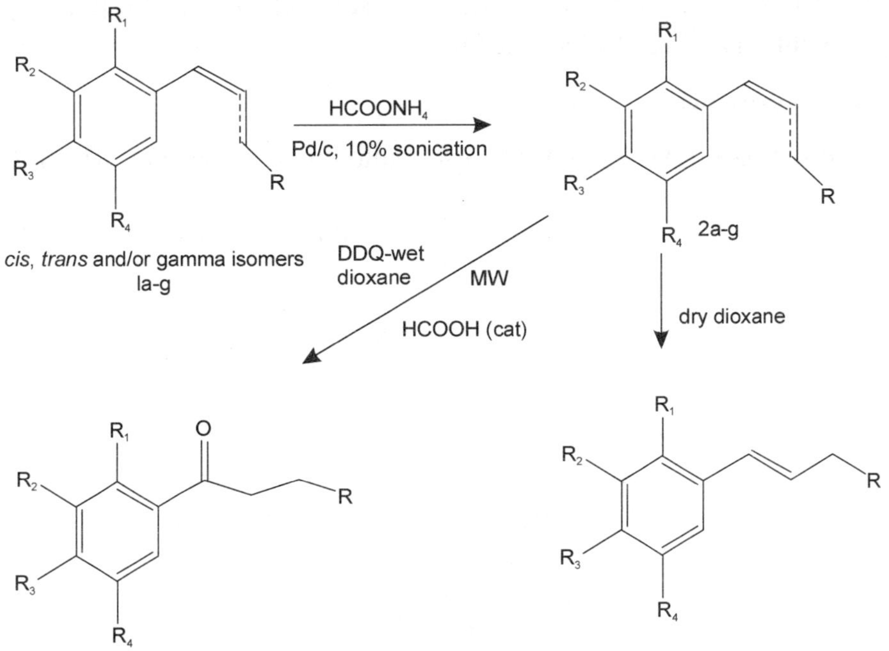

Figure 4.9 Green methodology in chalcone synthesis (Continues)

	R^1	R^2	R^3	R^4	R
(a)	OMe	H	OMe	OMe	H
(b)	H	OMe	OMe	OMe	H
(c)	H	OMe	OMe	H	H
(d)	H	$(R_2 + R_3)$	$-OCH_2O-$	H	H
(e)	H	H	OMe	H	H
(f)	OMe	H	OMe	OMe	CH_3
(g)	H	OMe	OMe	H	CH_3
(h)	OMe	H	OMe	OMe	H
(i)	H	OMe	OMe	OMe	H

Figure 4.9 Green methodology in chalcone synthesis

Phenyl propanoids which occur naturally have been reported to be hypolipidemic agents apart from being neuroleptic, anti-ulcerogenic and anti-inflammatory. An ultrasound-assisted[47] method has been reported for the conversion of toxic methoxylated *cis*-isomer of aryl alkenes into its hypolipidemic active *trans*-isomer. Treatment of the mixture of isomers with ammonium formate and 10% Pd/C gave aryl alkanes which upon oxidation with DDQ in anhydrous dioxane containing a little amount of silica gel, provided (E)-arylalkenes in 42–72% yield depending on the substituents attached at the aryl ring. The same method on addition of a few drops of water, provided hypolipidemic active aryl alkanones (59–65% yield) (Figure 4.9).

The "Green" part of chemistry in the abovesaid reaction is the use of sonication, microwave condition and recyclable reagents.

Biocatalysed Synthesis of Salidroside

Salidroside (Rhodioloside) is a glycosidic compound found in the plant *Rhodiola rosea* and

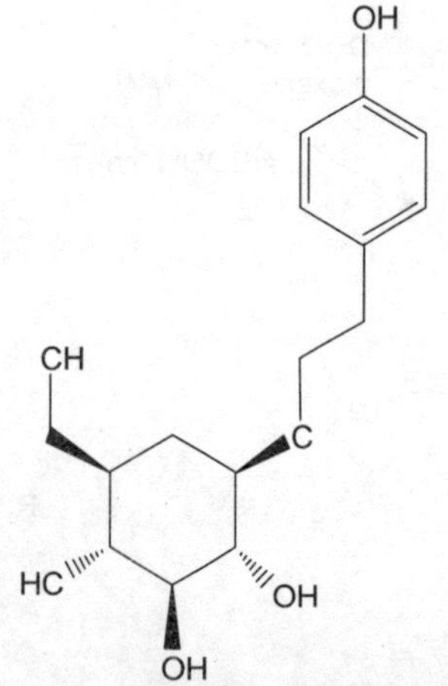

Salidroside

useful in resisting anoxia, microwave radiation and fatigue, improving oxygen lack, and postponing ageing. In recent years, salidroside has been used by divers, astronauts, pilots and mountaineers to enhance the ability for survival in adverse environment.

Since the *R. sachalinensis* A. Bor was an endangered species, multistep procedure including selective protection, activation and couplings were involved in preparing this in the laboratory scale.

But an environmentally benign synthesis of this glycoside using apple seed meal as green and robust biocatalyst has been reported.[48] (Figure 4.10).

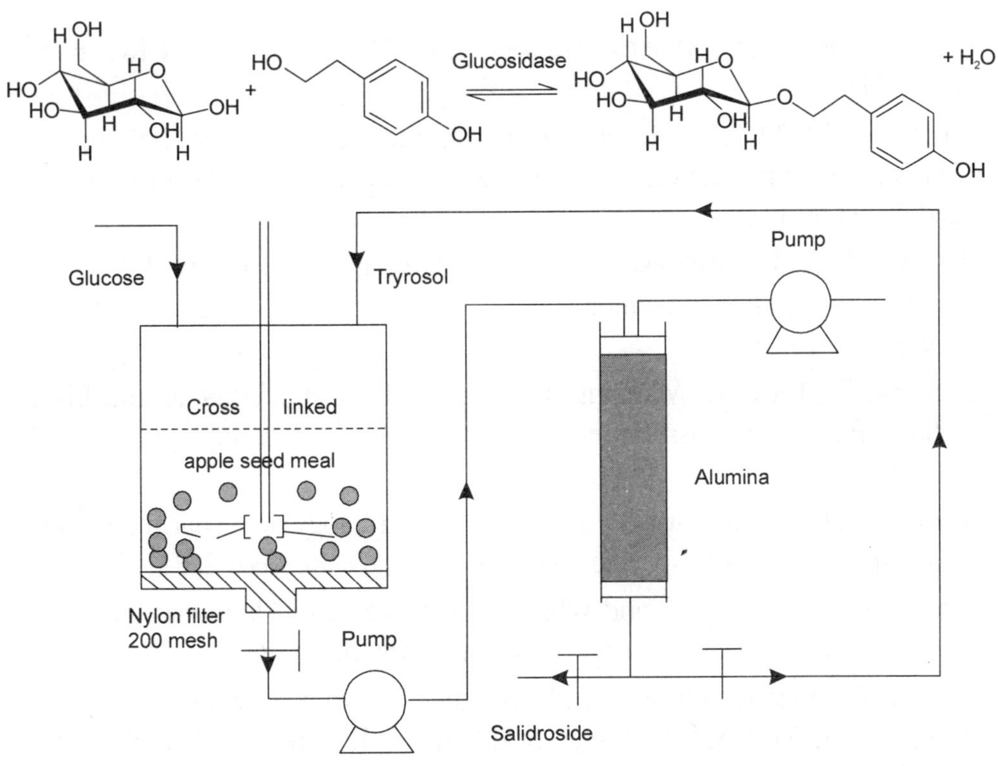

Figure 4.10 Enzymatic synthesis of salidroside catalysed by cross-linked apple seed meal integrated with alumina adsorption

The highlights of this enzyme-catalysed synthesis are

1. The whole process of production was environmentally benign and energy-economic.
2. The catalyst is from apple seed which is natural, easily available and also reusable.
3. The half-life of the native biocatalyst was enhanced up to 51 days via simple cross-linking immobilization.

4. The solvents used for reaction and extraction including t-BuOH, EtOH and EtOAc, were all clean, less toxic and recyclable.
5. The expensive substrate, tyrosol, can be recycled.

REFERENCES

1. Rice, E.L. (1984). *Allelopathy,* 2nd edn. *Academic Press,* New York. pp. 266–291.
2. Cornman, I. (1946). "Alterations on mitosis by coumarin and parascorbic acid." *Am.J.Bot.* 33, 217–219.
3. Putnam, A.R. (1985). *Weed Physiology,* Vol.1, Duke,S.O. (Ed.). CRC, Boca Raton. pp. 131–155.
4. Verheggen, M. Mescher, Haubruge, E., Moraes, C. and Schwartzberg, E. (2008). "Emission of alarm pheromone in aphids: a non-contagious phenomenon." *J. Chem. Ecol.* 34, 1146–1148.
5. Ray, A., Millar, J., Mcelfresh, J., Swift, I., Barbour, J. and Hanks, L. (2009). "Male-produced aggregation pheromone of the cerambycid beetle *Rosalia funebris.*" *J. Chem. Ecol.* 35, 96–103.
6. Suckling, D., Peck, R., Manning, L., Stringer, L., Cappadona, J. and El-Sayed, A. (2008). "Pheromone disruption of argentine ant trail integrity." *J. Chem. Ecol.* 34, p. 1609.
7. Nordlund, D. A. and Lewis, W. J. (1976). "Terminology of chemical releasing stimuli in intraspecific and interspecific interactions." *J. Chem. Ecol.* 2, 211–220.
8. Brown, W.L. Jr., Eisner, T. and Whittaker, R.H. (1970). "Allomones and kairomones: trans-specific chemical messengers." *Biosciences.* 20, 21–22.
9. Frey, M., Chomet, P., Glawischnig, E., Stettner, C., Grün, S., Winklmair, A., Eisenreich, W., Bacher, A., Meeley, R.B., Briggs, S. P., Simcox, K. and Gierl, A. (1997). "Analysis of a chemical defense mechanism in grasses." *Science.* 277: 696–699.
10. Wedge, D.E. and Camper, N.D. (2000). "Connections between agrochemicals and pharmaceuticals." In: *Biologically Active Natural Products: Agrochemicals and Pharmaceuticals.* Cutler, H.G. and Cutler, S.J. (Eds.). ACS Symposium Series, American Chemical Society, Washington DC.
11. Agrios, G.N. (1997). How pathogens attack plants. In: *Plant Pathology.* Academic Press, San Diego. 63–82.

12. Lee, R. and Balick, M.J. (2006). *"Chronobiology: It's about time." Explore.* 2(5), 442–445.

13. Dunlap, J.C. (1999). "Molecular bases for circadian clocks." *Cell.* 96:271–290.

14. Jouve, L., Gaspar, T., Kevers, C., Greppin, H. and Degli Agosti, R. (1999). "Involvement of indole-3-acetic acid in the circadian growth of the first internode of *Arabidopsis.*" *Planta.* 209, 136–142.

15. Stiebeling, B. and Neumann, K.H. (1986). "Identification and concentation of endogenous cytokinins in carrots (*Daucus carota L.*) as influenced by development and a circadian rhythm." *J. Plant Physiol.* 127, 111–121.

16. Novakova, M., Motyka, V., Dobrev, P.I., Malbeck, J., Gaudinova, A. and Vankova, R. (2005). "Diurnal variation of cytokinin, auxin and abscisic acid levels in tobacco leaves." *J. Exp. Bot.* 56, 2877–2883.

17. Novakova, M., Motyka, V., Dobrev, P.I., Malbeck, J., Gaudinova, A. and Vankova, R. (2005). "Diurnal variation of cytokinin, auxin and abscisic acid levels in tobacco leaves." *J. Exp. Bot.* 56, 2877–2883.

18. Salome, P.A., To, J.P.C., Kieber, J.J. and McClung, C.R. (2006). "*Arabidopsis* response regulators ARR3 and ARR4 play cytokinin-independent roles in the control of circadian period." *Plant Cell.* 18, 55–69.

19. Ruuhola, T., Tikkanen, O. and Tahvanainen, J. (2001). "Differences in host use efficiency of larvae of a generalist moth, *Operophtera brumata* on three chemically divergent Salix species." *J. Chem. Ecol.* 27, 1595–1615.

20. Saleem, A., Loponen, J., Pihlaja, K. and Oksanen, E. (2001). "Effects of long term open-field ozone exposure on leaf phenolics of European silver birch (*Betula pendula* Roth)." *J. Chem. Ecol.* 27, 1049–1062.

21. Giertych, M.J., Karolewski, P. and De Temmerman, L.O. (1999). "Foliage age and pollution alter content of phenolic compounds and chemical elements in *Pinus nigra* needles." *Water Air Soil Poll.* 110, 363–377.

22. Kause, A., Ossipov, V., Haukioja, E., Lempa, K. and Hanhimaki, S. (1999). "Multiplicity of biochemical factors of insect resistance in mountain birch." *Oecologia.* 120, 102–112.

23. Alonso, C., Ossipova, S. and Ossipov, V. (2002). "A high concentration of glucogallin, the common precursor of hydrolysable tannins, does not deter herbivores." *J. Insect Behav.* 15, 649–657.

24. Harborne, J.B. and Grayer, R.J. (1994). "Flavonoids and insects." In: Harborne, J.B. (Ed.). *The Flavonoids, Advances in Research Since 1986.* Chapman and Hall, London. pp. 589–618.

25. Simmonds, M.S.J. (2001). "Importance of flavonoids in insect–plant interactions: feeding and oviposition." *Phytochemistry.* 56, 245–252.

26. Blaney, W.M. and Simmonds, M.S.J. (1983). Electrophysiological activity in insects in response to antifeedants. COPR report Project 9; Overseas Development Organisation, London, UK.

27. Bernays, E.A., Oppenheim, S., Chapman, R.F., Kwon, H. and Gould, F. (1991). "Taste sensitivity of insect herbivores to deterrents is greater in specialists than in generalists: a behavioural test of the hypothesis with two closely related caterpillars." *J. Chem. Ecol.* 26, 547–563.

28. Bernays, E.A. and Chapman, R.F. (2000). "Plant secondary compounds and grasshoppers: beyond plant defences." *J. Chem. Ecol.* 26, 1773–1794.

29. Van Drongelen, W. (1979). "Contact chemoreception of host plant specific chemicals in larvae of various Yponomenta species (Lepidoptera)." *J. Comp. Physiol.* 134, 265–279.

30. Schoonhoven, L.M., Jermy, T. and van Loon, J.J.A. (1998). *Insect–Plant Biology: From Physiology to Evolution.* Chapman & Hall, London, UK.

31. Stonecipher, L.L., Hurley, P.S. and Netzly, D.H. (1993). "Effect of apigeninidin on the growth of selected bacteria." *J. Chem. Ecol.* 19, 1021–1027.

32. Tally, F. P. and DeBruin, M.F. (2000). "Development of daptomycin for gram-positive infections." *Journal of Antimicrobial Chemotherapy.* 46, 523–526.

33. Butler, M.S. (2005) "Natural products to drugs: natural product derived compounds in clinical trials." *Natural Product Reports.* 22, 162–195.

34. Goodman, J. and Walsh, V. (2001). *The Story of Taxol.* Cambridge University Press, Cambridge.

35. Holton, R.A., Somoza, C., Kim, H.B., Liang, F., Biediger, R.J., Boatman, P.D., Shindo, M., Smith, C.C. and Kim, S. *et al.* (1994). "First total synthesis of taxol. 1. Functionalization of the B ring." *Journal of the American Chemical Society.* 116, 1597–1598.

36. Holton, R.A., Kim, H.B., Somoza, C., Liang, F., Biediger, R.J., Boatman, P.D., Shindo, M., Smith, C.C. Kim, S. *et al.* (1994). "First total synthesis of taxol. 2. Completion of the C and D rings." *Journal of the American Chemical Society.* 116, 1599–1600.

37. Nicolaou, K.C., Yang, Z., Liu, J.J., Ueno, H., Nantermet, P.G., Guy, R.K., Claiborne, C.F., Renaud, J. Couladouros, E.A. *et al.* (1994). "Total synthesis of taxol." *Nature.* 367, 630–4.

38. Firn, R.D. and Jones, C.G. (2003). "Natural products—a simple model to explain chemical diversity." *Natural Product Reports.* 20, 382–91.

39. Pichersky, E., Sharkey, T.D. and Gershenzon, J. (2006). "Plant volatiles: a lack of function or a lack of knowledge?" *Trends in Plant Science.* 11,421.

40. Firn, R.D. and Jones, C.G. (2006). "Do we need hypothesis to explain VOC emissons." *Trends in Plant Science.* 11, 422.

41. Natsume, M. (1999). *Recent Res. Dev. Agric. Biol. Chem.* 3:11.

42. Natsume, M., (1999). "Differentiation of aerial mycelia–pamamycins and calcium ion in *Streptomyces alboniger.*" *Actinomyceto.* l 13:11.

43. Pogell, B.M. (1998). "The pamamycins: developmental autoregulators and antibiotics from *Streptomyces alboniger* A review and update." *Cell Mol. Biol.* 44, 461.

44. Germay, O., Kumar, N. and Thomas, E.J. (2001). "Total synthesis of pamamycin 607: Applications of remote asymmetric induction in organic synthesis." *Tetrahedron Lett.* 42, 4969.

45. a. Anastas, P.T. and Warner, J.C. (1998). *Green Chemistry: Theory and Practice*. Oxford University Press, New York.

    b. Mason, T.J. and Cintas, P. (2002). In: *Handbook of Green Chemistry and Technology.* Clark, J. and Macquarrie, D.(Eds). Blackwell Publishing , *London.*

46. Palleros, D.R. (2004). "Solvent-free synthesis of chalcones." *J. Chem. Edu.* 81, 1345.

47. Joshi, B.P., Sharma, A . and Sinha, A.K. (2005). "Ultrasound-assisted convenient synthesis of hypolipidemic active natural methoxylated (*E*)-arylalkenes and arylalkanones." *Tetrahedron.* 61, 3075.

48. Hui-Lei Yu, Jian-He Xua, Wen-Ya Lu and Guo-Qiang Lin (2008). "Environmentally benign synthesis of natural glycosides using apple seed meal as green and robust biocatalyst." *Journal of Biotechnology.* 133, 469–477.

5

Nutraceuticals

Even though medicinal and pharmaceutical sciences, through the development of technology, have created milestones, plant-based systems continue to play an essential role in the health care of many communities. It has been estimated by the World Health Organization that approximately 80% of the world's inhabitants rely mainly on traditional medicines for their primary health care.[1] For the remaining 20% of the world's population, mainly residing in developed countries, nature is equally important since approximately 25% of the prescribed drugs contain extracts or plant metabolites and an additional significant percentage of the market drugs have been developed through studies employing natural products as the lead molecules.[2]

With the increased rampancy in diseases, there is a great demand for newer drugs and more consciousness on prevention of diseases. The food that we consume, forms an essential part of certain nutritional factors that can prevent or retard the progress of disease. New concepts have appeared with this trend, such as nutraceuticals, nutritional therapy, phytonutrients and phytotherapy.[3–5]

These functional or medicinal foods and phytonutrients or phytomedicines play positive roles in maintaining the well-being, enhancing health and modulating immune function to prevent specific diseases. The term nutraceutical was coined by DeFelice, Director of the Foundation for Innovation in Medicine.[6,7]

By definition, a nutraceutical, a combination of the words nutrition and pharmaceuticals, is any substance that is a food or part of a food and provides medical or health benefits, including the prevention and/or treatment of disease. Such products may range from isolated nutrients, dietary supplements, and diets to genetically engineered designer foods, herbal products, and processed foods such as cereals, soups, and beverages. It applies to a bioengineered designer vegetable food rich in antioxidant ingredients as well as to a stimulant functional food or pharma food. From a regulatory point of view, the term *nutraceutical* includes both a dietary supplement as defined under the Dietary Supplement Health and Education Act (DSHEA) and a medical food as defined in the Orphan Drug Act. Recognizing the role of/use of enormous medical phytochemicals in health, one notable point is that though nutraceuticals are currently widely in vogue, their efficacy dosage as well as their nutrient–drug interaction is yet to be well-defined. The following are the three significant contributors for "Nutraceuticals" to gain prominence.

1. Modern drugs with limited success in treating many degenerative diseases, such as osteoarthritis, have triggered the exploration for alternatives with less harmful side- effects.
2. Phytotherapy, be it ayurveda, siddha or unani medicines, has been recorded as traditional folklore medicine.
3. Growing knowledge from nutritional research and pharmacological or epidemiological studies has suggested a close relationship between nutritional therapy or nutraceutical intake and the promotion of health conditions.

A recent clinical trial has shown the effectiveness of enteral supplement with glutamine, arginine, and omega-3 fatty acid on acute severe pancreatitis.[8]

CURRENT STATUS OF NUTRACEUTICALS/DIETARY SUPPLEMENTS

Table 5.1 lists some of the widely used nutraceuticals/dietary supplements used currently.

Table 5.1 Some nutraceuticals and their benefits

Nutraceutical	Benefit claimed	Bioactive component
Garlic pills	Boosts immunity, imparts cardiovascular strength	Allicin
Circumax	Cholesterol control, improved blood circulation, cardiovascular benefits	Combination of four herbs Garlic, Hawthorne berry, *Gingko biloba, Andrographis paniculata* and Asian Ginseng

(Contd.)

Table 5.1 (Continued)

Nutraceutical	Benefit claimed	Bioactive component
Spirulina	Anaemia control	Phycocycanin
Valerian	Insomnia, anxiety	Iridoid triesters
Glucosamine sulphate	Arthritis	Glucosamine
Resveratrol	Health heart, anti-parkinson	Polyphenol
Ashwagandha *(Withania somnifera)*	Stress buster	Withanolides
Ginseng	Male infertility, diabetes, epilepsy	Gingenosides

Various patents have been filed in this area and a few of them are given in Table 5.2.

Table 5.2 Some patents filed in nutraceuticals

Authors	Patent No.	Titles and main achievement
Fan *et al.*	US7049433	Glucosamine and method of making glucosamine from microbial biomass
Wolfram, Loon van	WO06077202A1 EP1633208A1	Novel nutraceutical compositions
Romero, T.Miller, P.	WO06020091A3 WO06020091A2	Dietary supplements containing extracts of cinnamon and methods of using the same to promote weight loss
Shen B.J., Ghosh, P.	AU20020112A4 ZA0406621A	Nutraceuticals for the treatment, protection and restoration of connective tissues
Raju, G.G.	US20050282894A1	Hydroxycitric acid compositions, pharmaceutical and dietary supplements, and food products made therefrom and methods for their use in reducing body weight
Boreyko B.K., Wang, Y.	US20050266018A1	Nutraceutical compositions with mangosteen
McKee, D., Karwic, A.	US20060134294A1	Product and method for oral administration of nutraceuticals

ETHNOBOTANY AND PHARMACOLOGY

Plant remedies, as with pharmaceuticals, may be used for general or specific indications. This may be the result of the active pharmacophores having general or specific mechanisms of action or more specifically, acting at early stages in event cascades, versus late stages. For example, inhibition of cyclooxygenase by salicylic acid (found in many plants) affects many systems. Because of its general anti-inflammatory activity, it can be used to treat many types of inflammation, and can also affect other apparatus since it is widely used to enhance the bloodstream and prevent heart attacks.

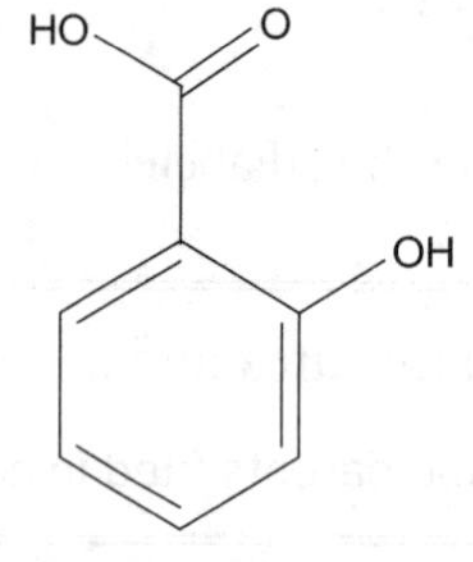

Salicylic acid

Colchicine (from autumn crocus, *Colchicum autumnale* L.) acts specifically as an anti-mitotic agent, binding tubulin.

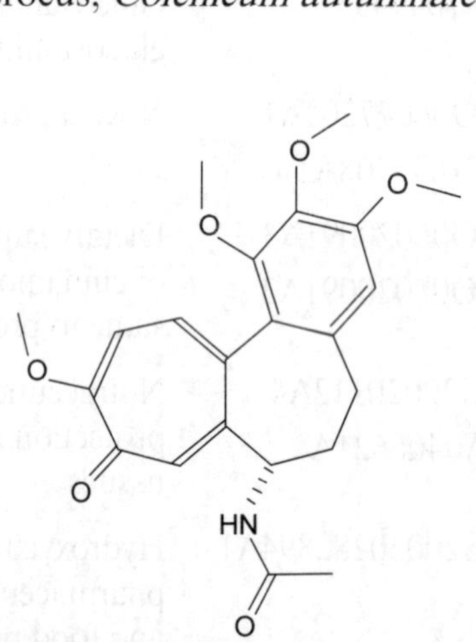

Colchicine

It, thereby, interferes with the function of mitotic spindles and thus causes de-polymerization and lack of motility in granulocytes and other motile cells active in the propagation of inflammatory processes such as gout. Although both drugs are useful in the treatment of inflammation, colchicine has specific activity, while salicylic acid is more general in its effects. There are also many mechanisms of CNS activity which have not yet been elucidated and for which ethnobotanical research may provide insight.

Ethnobotany and Natural Product Research

Ethnobotany, the largest subdiscipline of ethnobiology, is generally defined as the "science of people's interaction with plants" . This circumscription includes the study of plants that have therapeutic applications. While the primary objectives of modern ethnobotany are neither to develop new pharmaceuticals, nor to discover new bioactive chemical moieties, elucidating the pharmacological activities of a particular plant is part of some ethnobotanists' research.

Table 5.3 gives an idea about the disease conditions and their homologues in phytomedicine.

Table 5.3 Therapeutic indication and terminology

Therapeutic indication	The homolog of therapeutic indication in biomedicine
To fortify the brain	Adaptogen
For insomnia	Hypnotic
As a depurative	Blood thinner
For weight loss	Anoretic or thermogenic
As a sedative	Anxiolytic
For rejuvenation	Adaptogenic
For insanity	Neuroleptic
For headaches and body pain	Analgesic
To alter the mind	Hallucinogen
For sexual enhancement	Aphrodisiac
To energize	Tonic/Adaptogen
For muscle building	Anabolic
To stimulate appetite	Orexigenic

There is an open website called Duke's phytochemical and ethnobotanical databases (http:// www.ars-grin.gov/duke/) which contains extensive information about ethanobotanicals.

The databases are divided and subdivided into the following: plant searches (chemicals and activities in a particular plant; high concentration chemicals; chemicals with one activity; ethnobotanical uses); chemical searches (plants with a chosen chemical; activities of a chosen chemical); activity searches (plants with a specific activity; search for plants with several activities; chemicals with a specific activity; chemicals with a lethal dose value); and ethnobotany searches (ethnobotanical uses for a particular plant; plants with a particular ethnobotanical use).

Two polyphenols (resveratrol and the proanthocyanidins) that are present in grapes have generated much excitement in the past couple of years for those wanting to increase their lifespan and brain power.

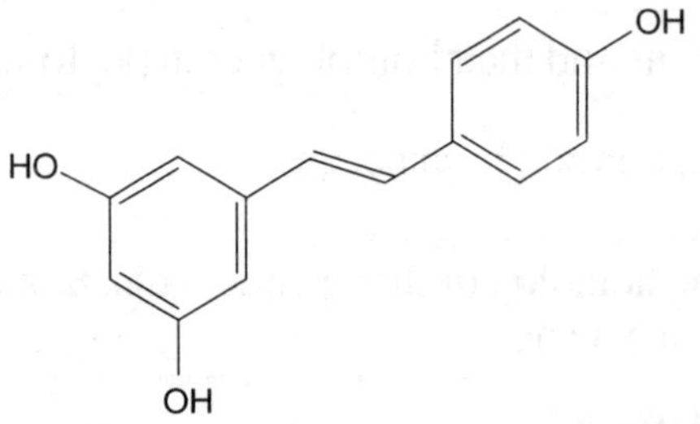

Resveratrol

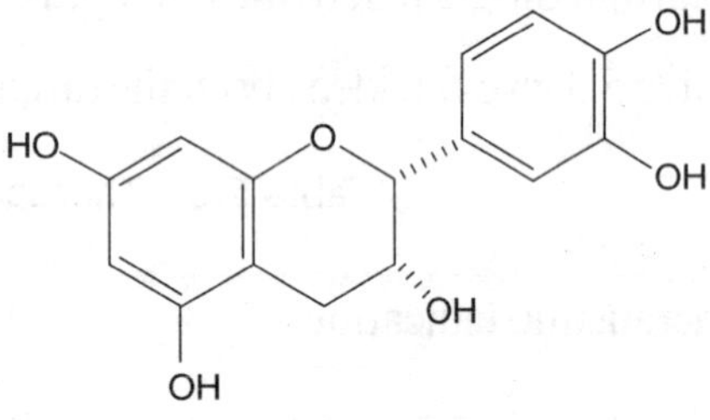

Proanthocyanidins

It has been long- known that a 30% reduction of normal calorie intake in mice causes them to live longer.[10] Rather than compounds from patented pharmaceutical libraries, the most active group is the polyphenols, in particular the grape polyphenol resveratrol (a stilbene).[11] When included in the yeast growth medium, resveratrol extended the life of wildtype yeast. Similar results were reported in *Drosophila* and *Caenorhabditis.*[12]

NUTRACEUTICAL ANTIOXIDANTS AND HEALTH

One of the main causes of various diseases are attributed to free radicals. Free radicals are reactive species which could damage a bioprocess in the human system. Certain reactive di-oxygen species include superoxide anion (O_2^{2-}), hydroxyl radical (OH), peroxy radical (ROO), nitric oxide (NO), etc. The reactivity of free radicals comes from the presence of at least one unpaired electron. They are formed as mediators or by-products of processes such as neurotransmission and inflammatory reactions in the human system apart from external causes such as pollution. Free radicals are involved in the pathology of a number of conditions. Oxidized low density lipoprotein (LDL) is involved in the development of atherosclerosis. Free radicals play several roles in the development of cancer. They contribute to the development of diabetes by playing a role in the autoimmune destruction of β cells while also impairing the action of insulin. Type I diabetes is caused by the destruction of the pancreatic beta cells responsible for producing insulin. In humans, the diabetogenic process appears to be caused by immune destruction of the beta cells; part of this process is apparently mediated by white cell production of active oxygen species. Diabetes can be produced in animals by the drugs alloxan and streptozotocin; the mechanism of action of these two drugs is different, but both result in the production of active oxygen species. Scavengers of oxygen radicals are effective in preventing diabetes in these animal models. Not only are oxygen radicals involved in the cause of diabetes, they also appear to play a role in some of the complications seen in long-term treatment of diabetes. Free radicals have even been implicated in asthma, rheumatoid arthritis, and cataracts.[13]

When the production of reactive species surpasses the body's endogenous antioxidant network's ability to provide protection, a state of "oxidative stress" occurs. With increasing age, the oxidation products from lipids, nucleic acids, proteins, sugars, and sterols are found to increase. The main causes of the aging process thus seems to be related to reactive oxygen species and free radicals, such as superoxide anion, hydrogen peroxide, hydroxyl radicals, and singlet oxygen. Mitochondria, which consume more than 90% of the oxygen in aerobic living organisms, are the main reactive oxygen species and free radical source. A serious imbalance between reactive oxygen species and antioxidants causes oxidative stress. Oxidative stress is caused by antioxidant-deficient diets or by increased production of reactive oxygen species by environmental toxins such as those caused by smoking or by inappropriate activation of phagocytes such as with chronic inflammatory disease. Clinical studies reported that reactive oxygen species are associated with many degenerative diseases, which are associated with aging (Figure 5.1).

As oxidative stress increases, the level of pro-oxidants against antioxidants increases and the aging process is accelerated. If reactive oxygen species and free radicals are the major causes of aging processes, antioxidative nutraceuticals can reduce the level of reactive oxygen species and free radicals, slow the aging process, and increase the lifespan. Antioxidants terminate free radical reactions by removing radical intermediates and inhibit other oxidation reactions by being oxidized themselves and thus becoming free radicals. The newly created free radicals are relatively weak and therefore are not likely to do further harm. Antioxidative nutraceuticals can inhibit or slow the formation of free alkyl radicals in the initiation step and interrupt the free-radical chain reactions in the propagation step during lipid oxidation. Antioxidants can "quench," or stop, free radicals before they cause harm, much like soaking apple slices in lemon juice (that contain high levels of ascorbic acid or vitamin C) preventing the oxidative reaction that occurs when food is exposed to air. Antioxidative nutraceuticals can be antioxidative enzymes, hydrogen-donating compounds, metal chelators, and singlet oxygen quenchers. It has been reported that levels and activities of antioxidant enzymes, including superoxide dismutase, catalase, and glutathione peroxidase, are much higher in long-living species than in short-living ones. The amount of vitamin E in the human system has been found to be inversely proportional to age and quantified vitamin A and E have been reported to increase the life expectancy in animals.[14, 15]

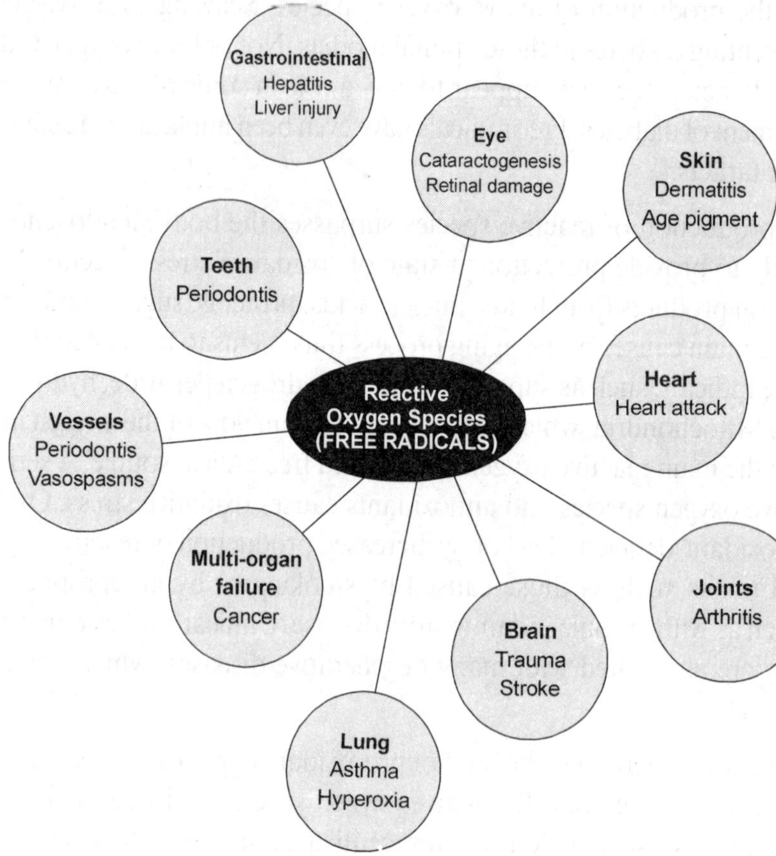

Figure 5.1 Clinical conditions involving reactive oxygen species

Apart from these there are some enzymes which possess antioxidant property, for example, superoxide dismutase (for superoxide removal), catalase (hydroperoxide removal), glutathione peroxidase (oxidized glutathione reduction), glutathione S-transferase (lipid hydroperoxide removal), methionine sulphoxide reductase (repair of oxidized methionine residues), etc.

Coenzyme Q10

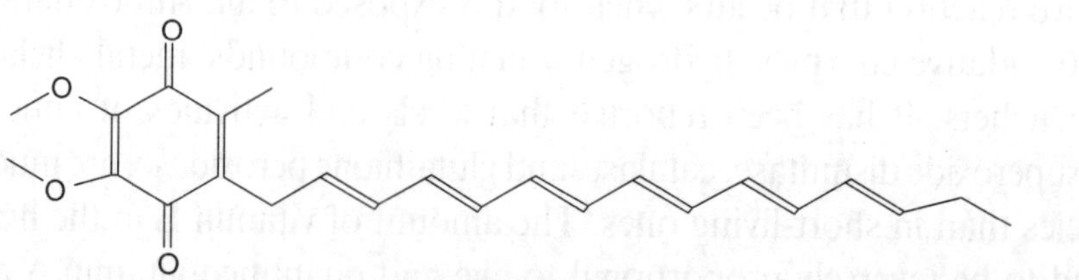

Ubiquinone

Ubiquinone, more commonly known as coenzyme Q10, is a substance found in the body as a component of the mitochondrial respiratory chain. It works in concert with other substances to regenerate ATP (energy) in a cell. Coenzyme Q10 also functions as a powerful antioxidant and free-radical scavenger. An antioxidant is a substance that gives up electrons easily and can act to neutralize, harmful oxidants and free-radicals. In humans, the levels of coenzyme Q10 have been found to be below the normal, in patients with cardiovascular disease and periodontal disease. Whether low levels are a cause or effect is not clear, but coenzyme Q10 supplementation has been reported to have been used successfully in the treatment of heart problems, muscular dystrophy, myopathies and periodontal disease.[16, 17]

Ascorbic acid

Vitamin C or ascorbic acid is reported to have preventive effects on some cancers, heart disease, and the common cold. Ascorbic acid and tocopherol supplementation can substantially reduce oxidative damage. The effect is substantiative in nonsmokers than smokers, since smoking induces oxidative stress, which form free-radical compounds in the gas phases and the ascorbic acid radical could be pro-oxidant in smokers.[18] The antioxidant mechanisms of ascorbic acid are based on hydrogen atom donation to lipid radicals, quenching of singlet oxygen, and removal of molecular oxygen (Figure 5.2).

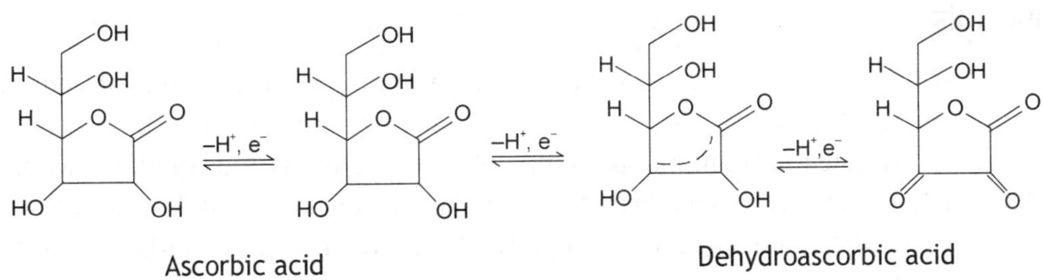

Figure 5.2 Sequential one-electron oxidation of L-ascorbic acid

Carotenoids

Carotenoids are a group of tetraterpenoids. The basic carotenoid-structural backbone consists of isoprenoid units formed either by head-to-tail or by tail-to-tail biosynthesis. There are primarily two classes of carotenoids: carotenes and xanthophylls (Figure 5.3) .The antioxidant potentials of carotenoids have been reported for the prevention of free-radical-initiated diseases, including atherosclerosis, cataracts, age-related muscular degeneration, and multiple sclerosis. Lycopene, which is the main carotenoid of tomatoes and tomato products, has several health benefits including decreasing the development of cervical, colon, prostate, rectal, stomach, and other types of cancers.[19, 20]

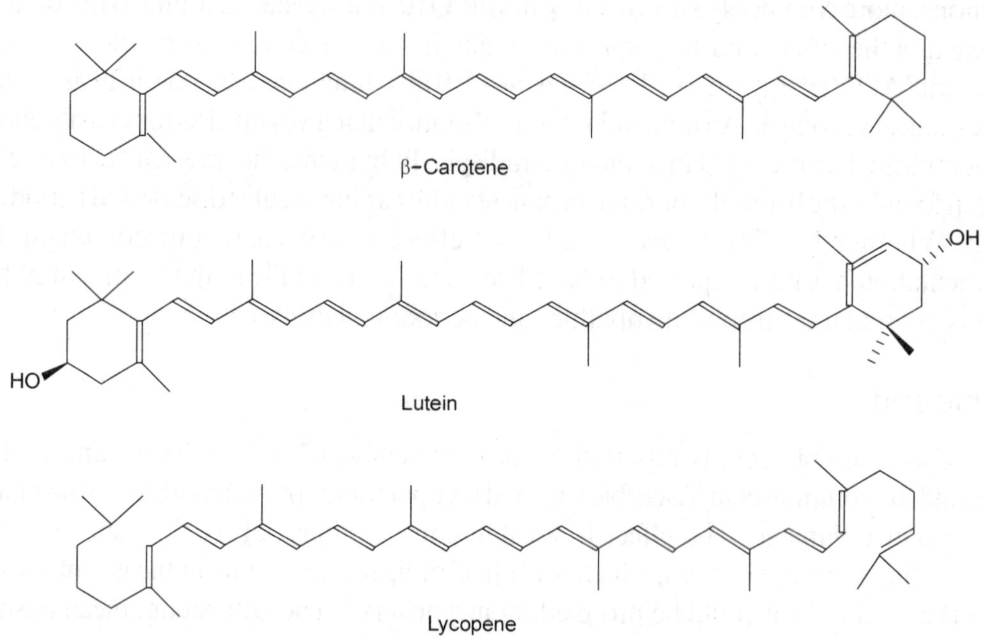

Figure 5.3 Some carotenoids

Polyphenols

Polyphenols are phenolic compounds possessing antioxidant properties, found abundantly in berries, broccoli, red wine, chocolate, green tea, etc. Their classification based on structure is given in Figure 5.4. Phenolic compounds or polyphenols are ubiquitous in plants with more than 8000 structures reported.[25] It has been reported that phenolic compounds have antioxidant, antimutagenic, and free-radical-scavenging activities. Epidemiological studies showed that increased consumption of phenolic compounds reduces the risk of cardiovascular disease and certain types of cancer. Moderate consumption of red wine, which contains high content of polyphenols, is associated with a low risk of coronary heart disease.[25] Consumption of soy and soy products are related with biological effects, including anticarcinogenic, antiatherosclerotic, and antihaemolytic effects. Soybean is the unique source of isoflavones with 1–3 mg/g and with 0.025–3 mg/g soy products.[26] Antioxidant activities of isoflavones, especially genistein, were reported *in vivo* and *in vitro*, in simple lipid system such as liposomes, and in more complex systems such as lipoproteins. Addition of purified forms of isoflavones inhibited copper-dependent LDL oxidation.[27] Oral intake of the isoflavone genistein is associated with an increased resistance of LDL oxidation and inhibition of plasma lipid oxidation products.[28] Antioxidant mechanisms of polyphenolic compounds are based on hydrogen donation abilities and chelating metal ions.

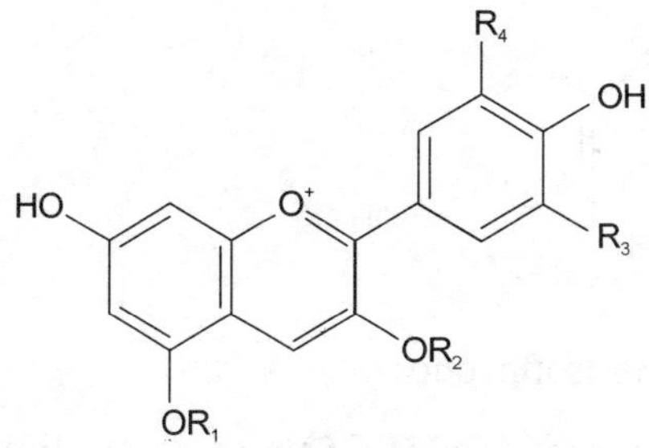

Anthocyanins

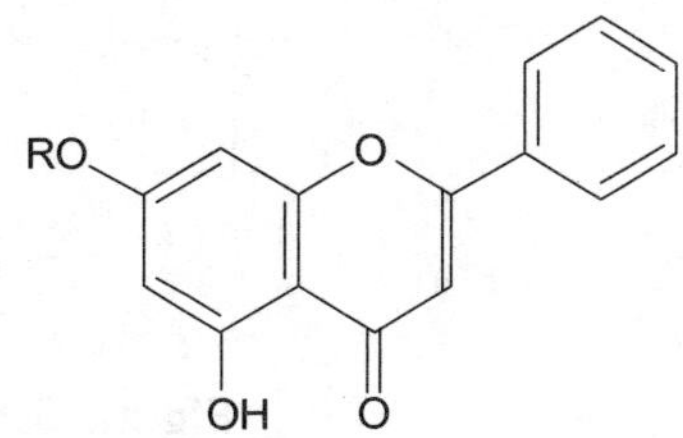

Flavones

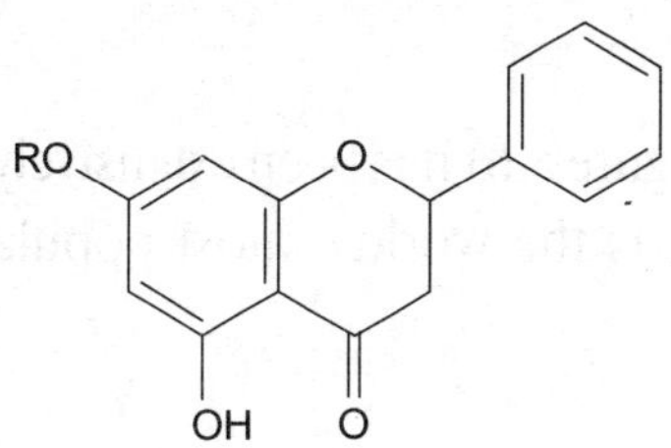

Flavanones

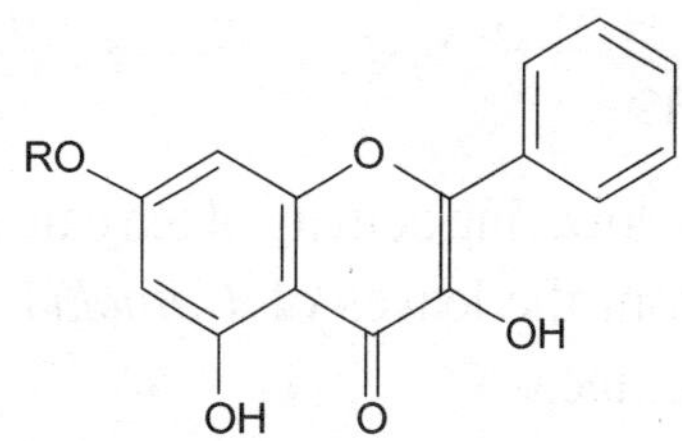

Flavanols

Figure 5.4 Chemically varying flavonoids

Genetically modified tomatoes contain high levels of flavonols such as quercetin, kaempferol, and glycosides and flavones such as luteolin, lycopene and luteolin 7-glucoside in their peel tissue.[21,22] Resveratrol level in *Brassica napus* seed has also dramatically increased.[23] Consumption of these transgenic tomatoes has been shown to yield certain health benefits in mice.[24]

Isoflavones

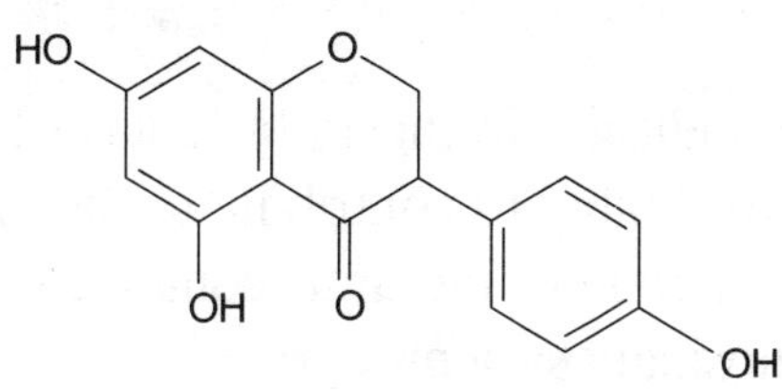

Genistein

Daidzein

Figure 5.5 Some isoflavones *(Continues)*

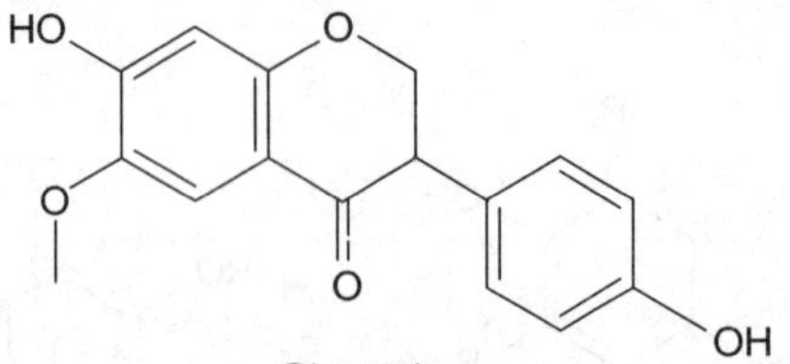

Figure 5.5 Some isoflavones

The structural similarity between the isoflavones(Figure 5.5) and oestradiol is the basis for the proposition that the isoflavones may be able to replace human oestrogenic activity.

Tea Catechins

The cholesterol-lowering activity of tea catechins (Figure 5.6) has been extensively investigated. Tea, derived from the leaves of *Camellia sinensis*, is the world's most popular and widely consumed beverage.

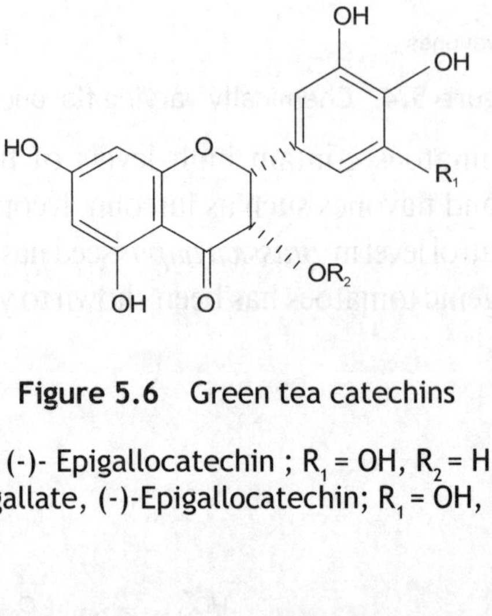

Figure 5.6 Green tea catechins

(-)- Epicatechin; $R_1 = R_2 = H$ (-)- Epigallocatechin ; $R_1 = OH$, $R_2 = H$, (-)- epicatechin gallate; $R_1 = H$, R_2 = gallate, (-)-Epigallocatechin; $R_1 = OH$, R_2 = gallate

Quercetin

Quercetin is the strongest of the flavonoid free-radical scavengers (it contains five hydroxyl groups) and it is suggested that quercetin has the ability to contribute to the total plasma antioxidant capacity more than six times that of vitamin C and urate combined.[29] Quercetin also has the ability to boost the endogenous antioxidant system.

Quercetin is a very efficient free-radical scavenger. It enhances endurance and performance. As research demonstrates that radicals and other reactive oxygen species (ROS) are an underlying aetiology in exercise-induced physiological disturbances.[30,31] It is known to possess anti-inflammatory activity,[32] positive psycho-stimulant effects,[33] and, especially, mitochondrial biogenesis in the muscle and brain of mice[34]. Quercetin has also demonstrated other biological effects in preclinical testing, including anti-inflammatory, antiviral, and anticancer activity, as well as protective effects on the heart and nervous system.[35, 36]

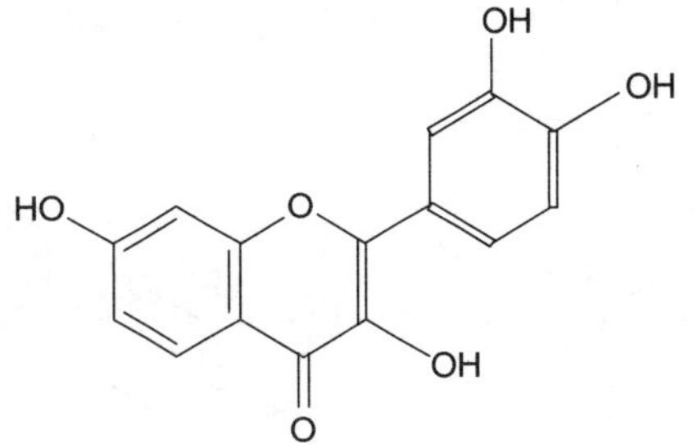

3′, 4′, 5′, 7-pentahydroxyflavone (Quercetin)

Phytosterols

Plant sterols and their saturated derivatives, stanols, are a group of cholesterol analogues with different side-chain configurations (Figure 5.7). Phytosterols are plant analogues of mammalian cholesterol. The principal sterols are sitosterol, campesterol, and stigmasterol. These phytosterols have been sold as functional cholesterol-lowering nutraceuticals in Europe, the United States, and Australia.

A major application for these phytosterols is their addition into spreads and vegetable oils (functional margarine, butter, and cooking oils). It is estimated that the phytosterol intake in humans can reach 160–360 mg/day, and it has been suggested that the daily consumption of 2 g of phytosterols can effectively lower the cholesterol by 9–14% in humans with little or no effect on HDL-C and TG levels.

Besides resisting absorption themselves, phytosterols are also inhibitors of intestinal cholesterol absorption, thereby lowering plasma cholesterol levels.[37]

First, phytosterols may cause an effective displacement of cholesterol from micellar binding in the intestine and therefore diminish cholesterol absorption.[38]

Second, phytosterols may affect cholesterol synthesis, as sitosterol has been shown to decrease cholesterol synthesis by inhibiting HMG-CoA reductase gene expression in CaCo–2 cells.[39]

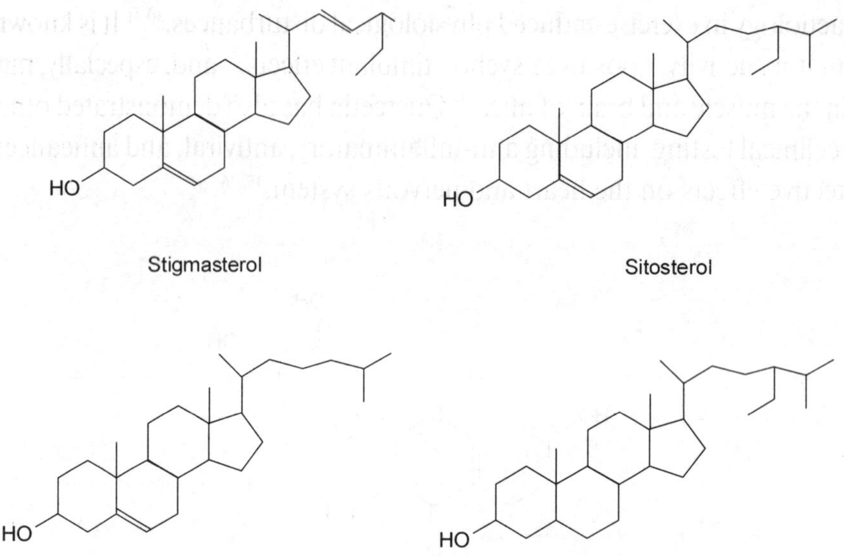

Figure 5.7 Plant sterols and stannols

Lipoic Acid

Some sulphur-containing compounds, including glutathione, lipoic acid, and dihydrolipoic acid, have shown antioxidant activities. The chemical structure of lipoic acid is 1,2-dithilane-3-pentanoic acid.

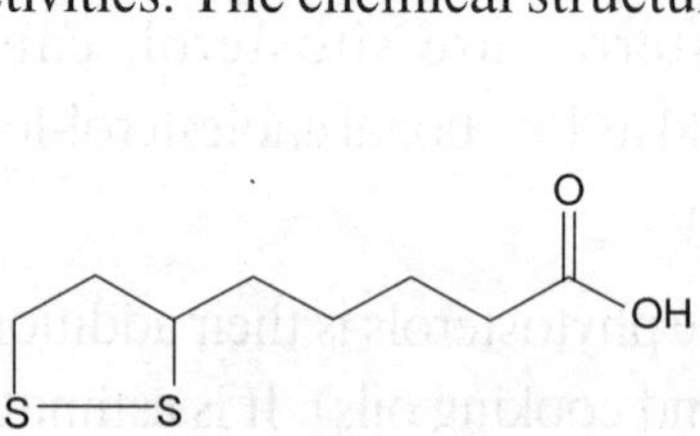

Lipoic acid

Lipoic acid is present in meat, liver, and heart. Lipoic acids can prevent oxidative damages of proteins. Antioxidant activity of lipoic acid can help to reduce diabetic relative complications, which can be developed through oxidative stress. Lipoic acid plays an important role in reducing blood glucose concentration. Lipoic acid regenerates GSH in liver, kidney, and lung-tissue and also regenerates vitamins C and E. A dietary study of lipoic acid showed a decrease in age-related decline in oxygen consumption and radical formation, improvement of mitochondrial membrane potential, and increase in ascorbic acid and GSH levels.[40] Lipoic acids may improve age-related decline in memory and cognitive function and brain-related ailments, including Alzheimer's disease and Parkinson's disease.[41] Reduced (dihydrolipoic acid) and oxidized forms of lipoic acid both act as antioxidants and scavenge the reactive oxygen species. Lipoic acids are excellent antioxidants, showing abilities for radical scavenging, metal chelating, interaction with other antioxidants, metabolic regeneration, and gene regulation.[42]

Curcumin

Curcumin is the main component of *Curcuma longa* that is traditionally used in cooking.

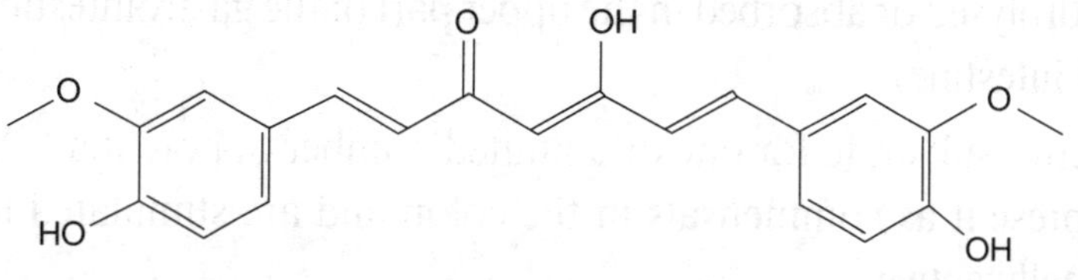

Curcumin

This phenolic compound is insoluble in water under acidic or neutral conditions but dissolves in alkaline conditions. It has been reported to possess anti-cancer,[43-45] antioxidant,[46, 47] hyperlipidaemic[48, 49] and hepatoprotective[50, 51] activities apart from several other pharmcological action. It has been reported to possess insulin-sensitization ability and hence may be of use in diabetes. Both NF–κ B and TNF have been linked with the induction of resistance to insulin. Because curcumin can down-regulate the activation of NF–κ B and down-regulate TNF expression and TNF signalling, it can be exploited in diabetic patients. Several animal studies have demonstrated that curcumin can overcome insulin resistance.[52–54]

Chlorogenic Acid

Chlorogenic acid is an ester of caffeic and quinic acid and is found abundantly in coffee. Being a polyphenol, it functions as an antioxidant, and *in vitro* studies shows that it is a free- radical scavenger.[55] It has also been reported to increase the resistance to lipid peroxidation of LDL. It has been shown to inhibit carcinogenesis in small animals.[56]

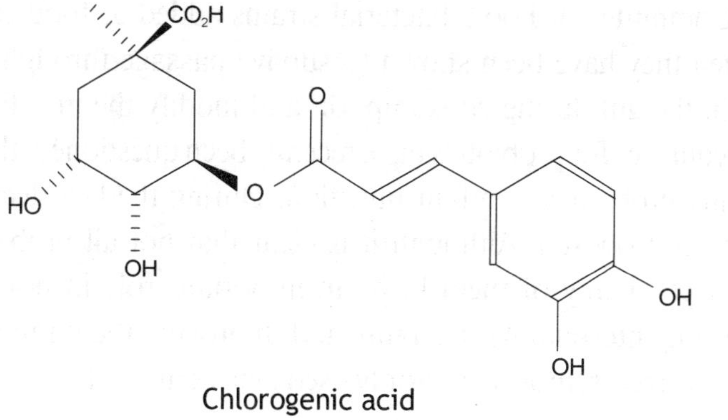

Chlorogenic acid

Prebiotics and Probiotics

Prebiotics are defined as non-digestible food ingredients that beneficially affect the host by selectively stimulating the growth and/or activity of one or a limited number of bacteria in the colon, and thus improve host health.

In order to be classified as a prebiotic, a food ingredient should

1. not be hydrolysed or absorbed in the upper part of the gastrointestinal tract (including the small intestine)
2. be a selective substrate for one or a limited number of potentially beneficial bacteria that are present as commensals in the colon and are stimulated to grow and/or be metabolically active
3. be able to alter the colonic flora in favour of a potentially more healthy composition and
4. induce luminal or systemic effects that are beneficial to the host health.

Several food ingredients like non-digestible carbohydrates, some peptides, proteins and lipids, have been proposed as prebiotics. Although all non-digestible carbohydrates are fermented by intestinal microflora, not all of them can be classified as prebiotics because they have been found to also stimulate the growth and/or activity of potentially harmful bacteria. Fructo-oligosaccharides (FOS) and inulin are prebiotics, naturally present in plants such as onion, chicory and asparagus (18). Inulin and FOS are mostly linear polymers of fructose with glucose as the terminal sugar. FOS have a degree of polymerization (DP), i. e., number of monosaccharide units, between 2 and 7, whereas inulin has a DP of 2 to 60.

Probiotics are originally defined as live microbial food supplements which beneficially affects the host animal by improving its intestinal microbial balance. Live microorganisms which when administered in adequate amounts confer a health benefit on the host."According to the Scientific Committee of Food, bacterial strains added to food can be considered as generally safe when they have been shown to survive passage through the gastrointestinal tract, proliferate in the gut during consumption and modify the intestinal milieu, e.g., pH. Although the prerequisites for probiotics have recently been questioned, there is a considerable interest in including probiotics in infant nutrition. During the last decade, many potential probiotics have been proposed. Although it is clear that not all probiotics have the same positive health effects, many of them have an important role in normalizing the altered microflora, increasing intestinal permeability and improving the immune barrier functions. Among the probiotic agents, most commonly used genera are lactobacilli and bifidobacteria.

REFERENCES

1. Farnsworth, N.R. *et al.* (1985). "Medicinal plants in therapy." *Bull. World Health Organ.* 63, 965–981.
2. Cragg, G.M. and Newman, D.J. (2005). "International collaboration in drug discovery and development from natural sources." *Pure Appl. Chem.* 77, 1923–1942.
3. Bland, J.S. (1996). "Phytonutrition, phytotherapy, and phytopharmacology." *Altern. Ther. Health Med.* 2, 73–76.
4. Berger, M.M. and Shenkin, A. (2006). "Vitamins and trace elements: Practical aspects of supplementation." *Nutrition.* 22, 952–55.
5. Bagchi, D. (2006). "Nutraceuticals and functional foods regulations in the United States and around the world." *Toxicol.* 221, 1–3.
6. Anonymous. (1991). The nutraceutical initiative: a proposal for economic and regulatory reform (white paper), Foundation for Innovative Medicine, New York, USA.
7. DeFelice, S.L. (1995). "The nutraceutical revolution: its impact on food industry R&D." *Trends Food Sci. Technol.* 6, 59–61.
8. Pearce, C.B., Sadek, S.A. and Walters, A.M. *et al.* (2006). "A double-blind, randomised, controlled trial to study the effects of an enteral feed supplemented with glutamine, arginine, and omega-3 fatty acid in predicted acute severe pancreatitis." *JOP.* 7, 361–371.
9. Bertelli, A.A. and Das, D.K. (2009). "Grapes, wines, resveratrol, and heart health, *J. Cardiovasc. Pharmacol.* 54(6), 468–76.
10. McCay, C.M., Crowel, M.F. and Maynard, L.A. (1935). "The effect of retarded growth upon the length of the life span and upon the ultimatebody size." *J. Nutr.* 10, 63–79.
11. Horwitz, K.T., Bitterman, K.J., Cohen, H.Y., Lamming, D.W., Lavu, S., Wood, J.G., Zipkin, R.E., Chung, P., Kisielewski, A.and Zhang, L.L. *et al.* (2003). "Small molecule activators of sirtuins extend *Saccharomyces cerevisiae* lifespan." *Nature.* 425, 191–196.
12. Bauer, J.H., Goupil, S., Garber, G.B.and Helfand, S.L. (2004). "An accelerated assay for the identification of lifespan-extending interventions in *Drosophila melanogaster.*" *Proc. Natl. Acad. Sci.,* USA. 101, 12980–12985.
13. Knekt, P., Kumpulainen, J., Jarvinen, R., Rissanen, H., Heliovaara, M. and Reunanen, A., *et al.* (2002). "Flavonoid intake and risk of chronic diseases." *Am. J. Clin. Nutr.* 76(3), 560–568.

14. Duthie, S.J., Ma, A., Ross, M.A.and Collins, A.R. (1996). "Antioxidant supplementation decreases oxidative DNA damage in human lymphocytes." *Cancer Res.* 56, 1291–5.

15. Teoh, C.Y. and Davies, K.J.A. (2002). "The broad spectrum of antioxidant defense and oxidant repair mechanisms." *Free Rad. Res.* 36, 8–12.

16. Greenburg, S. and Frishman, W.H. (1990). "Coenzyme Q10: "A new drug for cardiovascular disease." *J. Clin. Pharmacol.* 30, 596–608.

17. Hishikawa, Y., Takahashe, M. and Yorifuji, S. (1989). "Long term coenzyme Q10 therapy for mitochondrial encephalomyopathy with cytochrome *c* oxidase deficiency." *Neurology.* 39, 399–403.

18. Kaur, C. and Kapoor, H.C. (2001). "Antioxidants in fruits and vegetables - the millennium's health." *Intl. J. Food. Sci. Tech.* 36, 703–25.

19. Giovannucci, E. and Clinton, S.K. (1998). "Tomatoes, lycopene, and prostate cancer." *Prostate cancer Proc. Soc. Exp. Biol. Med.* 218, 129–39.

20. Giovannucci, E. (1999). "Tomatoes, tomato-based products, lycopene, and cancer: Review of the epidemiologic literature." *J. Natl. Cancer Inst.* 91(4), 317–31.

21. Mehta, R.A., Cassol, T., Li, N., Ali, N., Handa, A.K. and Mattoo, A.K. (2002). "Engineered polyamine accumulation in tomato enhances phytonutrient content, juice quality, and vine life." *Nat. Biotechnol.* 206, 613–18.

22. Schijlen, E.G., de Vos, R., Jonker, H. *et al.* (2006). "Pathway engineering for healthy phytochemicals leading to the production of novel flavonoids in tomato fruit." *Plant Biotechnol. J.* 4, 433–44.

23. Hüsken, A., Baumert, A., Milkowski, C., Becker, H.C., Strack, D. and Möllers, C. (2005). "Resveratrol glucoside (Piceid) synthesis in seeds of transgenic oilseed rape (*Brassica napus* L.)." *Theor. Appl. Genet.* 111, 1553–62.

24. Rein, D., Schijlen, E., Kooistra, T. *et al.* (2006). "Transgenic flavonoid tomato intake reduces C- reactive protein in human C-reactive protein transgenic mice more than wild-type tomato." *J. Nutr.* 136, 2331–37.

25. Bravo, L. (1998). "Polyphenols: chemistry, dietary sources, metabolism, and nutritional significance." *Nutr. Rev.* 56, 317–33.

26. Wang, H.J. and Murphy, P.A. (1994). "Isoflavone content in commercial soybean foods." *J. Agri. Food Chem.* 42, 1666–73.

27. Hwang, J., Sevanian, A., Hodis, H.N. and Ursini, F. (2000). "Synergistic inhibition of LDL oxidation by phytoestrogens and ascorbic acid." *Free Rad. Biol. Med.* 29, 79–89.

28. Wiseman, H., O'Reilly, J.D., Adlercreutz, H., Mallet, A.I., Bowey, E.A., Rowland, I.R. and Sanders, T.A.B. (2000). "Isoflavone phytoestrogens consumed in soy decrease F2-isoprostane concentrations and increase resistance of low-density lipoprotein to oxidation in humans." *Am. J. Clinic. Nutr.* 72, 395–400.
29. Arts, M.J.T.J., Sebastiaan Dallinga, J., Voss, H.P., Haenen, G.R.M.M., and Bast, A. (2004). "A new approach to assess the total antioxidant capacity using the TEAC assay." *Food Chemistry.* 88(4), 567–570.
30. Powers, S.K., DeRuisseau, K.C., Quindry, J.and Hamilton, K.L. (2004). "Dietary antioxidants and exercise." *J. Sports Sci.* 22, 81–94.
31. Reid, M.B. (2008). "Free radicals and muscle fatigue: of ROS, canaries, and the IOC." *Free Radic. Biol. Med.* 44, 169–179.
32. Harwood, M., Danielewska-Nikiel, B. and Borzelleca, J.F., *et al.* (2007). "A critical review of the data related to the safety of quercetin and lack of evidence of *in vivo* toxicity, including lack of genotoxic/carcinogenic properties." *Food Chem. Toxicol.* 45, 2179–2205.
33. Alexander, S.P. (2006). "Flavonoids as antagonists at A1 adenosine receptors." *Phytother. Res.* 20, 1009–1012.
34. Davis, J.M., Murphy, E.A. and Carmichael, M.D. (2009). "Effects of the dietary flavonoid quercetin upon performance and health." *Curr. Sports Med. Rep.* 8, 206–213.
35. Ansari, M.A., Abdul, H.M., Joshi, G., *et al.* (2008). "Protective effect of quercetin in primary neurons against Abeta (1–42): relevance to Alzheimer's disease." *J. Nutri. Biochem.* 20(4), 269–75.
36. Utesch, D., Feige, K., Dasenbrock, J. *et al.* (2008). "Evaluation of the potential *in vivo* genotoxicity of quercetin." *Mutat. Res.* 654, 38–44.
37. John, S., Sorokin, A.V. and Thompson, P.D. (2007). "Phytosterols and vascular disease." *Curr. Opin. Lipidol.* 18, 35–40.
38. Heinemann, T., Kullak-Ublick, G.A., Pietruck, B. and vonBergmann, K. (1991). "Mechanisms of action of plant sterols on inhibition of cholesterol absorption. Comparison of sitosterol and sitostanol." *Eur. J. Clin. Pharmacol.* 40 (1), 59–63.
39. Field, F.J., Born, E. and Mathur, S.N. (1997). Effect of micellar β-sitosterol on cholesterol metabolism in Caco-2 cells. *J. Lipid Res.* 38, 348–360.
40. Hagen, T.M., Ingersoll, R.T., Lykkesfeldt, J., Liu, J., Wehr, C.M., Vinarsky, V., Bartholomew, J.C. and Ames, B.N. (1999). "(R)-lipoic acid-supplemented old rats

have improved mitochondrial function, decreased oxidative damage, and increased metabolic rate." *FASEB J.* 13(2), 411–8.

41. Kramer, K. and Packer, L. (2001). "R-lipoic acid." In: Kramer, K., Hoppe, P.P. and Packer, L., Eds. *Nutraceuticals in Health and Disease Preventi n.* Marcel Dekker, New York. pp. 129–64.

42. Bast, A. and Haenen, G.R.M.M. (2001). "Lipoic acid: a multifunctional nutraceutical. In: Kramer, K., Hoppe, P. P. and Packer, L. (Eds.). *Nutraceuticals in Health and Disease Prevention.* Marcel Dekker Inc. New York, pp. 113–28.

43. Ruby, A.J., Kuttan, G., Babu, K.D., Rajasekharan, K.N. and Kuttan, R. (1995). "Antitumor and antioxidant activity of natural curcuminoids." *Cancer Lett.* 94, 79–83.

44. Huang, M.T., Lou, Y.R., Xie, J.G., Ma, W., Lu, Y.P., Yan, P., Zhu, B.T., Newmark, H. and Ho, C.T. (1998). "Effect of dietary curcumin and dibenzoyl methane on formation of 7,12-dimethylbenz[a]anthracene induced mammary tumors and lymphomas (leukemia) in sencar mice." *Carcinogenesis.* 19, 1697–1700.

45. Gescher, A.J., Sharma, R.A. and Steward, W.P. (1992). "Cancer chemoprevention by dietary constituents: a tale of failure and promise." *Lancet Oncol.* 11, 226–230.

46. Shao, Z.M., Shen, Z.Z., Liu, C.H., Sartippour, M.R., Go, V.L., Heber, D. and Nguyen, M. (2002). "Curcumin exerts multiple suppressive effect on human breast carcinoma cells." *Int. J. Cancer.* 98, 234–240.

47. Sreejayan, N. and Rao, M.N.A. (1994). "Curcuminoid as potent inhibitor of lipid peroxidation." *J. Pharm. Pharmacol.* 46, 1013–1016.

48. Jayaprakasha, G.K., Jaganmohan, L. and Sakariah, K.K. (2006). "Antioxidant activities of curcumin, demethoxy curcumin and bisdemethoxy curcumin." *Food Chem.* 98, 720–724.

49. Rao, D.S., Sekhara, N.C., Satyanarayana, M.N. and Srinivasan, M. (1970). Effect of curcumin on serum and liver cholesterol level in rats." *J. Nutr.* 100, 1307–1315.

50. Patil, T.N. and Srinivasan, M. (1971). "Hypocholesteremic effect of curcumin in induced-hypercholesteremic rats." *Indian J. Exp. Biol.* 9, 167–169.

51. Despande, U.R., Gadre, S.G., Raste, A.S., Pillai, D., Bhide, S.V. and Samuel, A.M. (1998). "Protective effect of turmeric (*Curcuma longa* L.) extract on carbon tetrachloride-induced liver damage in rats." *Indian J. Exp. Biol.* 36, 573–577.

52. Babu, P.S. and Srinivasan, K. (1995). "Influence of dietary curcumin and cholesterol on the progression of experimentally induced diabetes in albino rat." *Mol. Cell Biochem.* 152, 13–21.

53. Arun, N. and Nalini, N. (2002). "Efficacy of turmeric on blood sugar and polyol pathway in diabetic albino rats." *Plant Foods Hum. Nutr.* 57, 41–52.

54. Matsuda, H., Tewtrakul, S., Morikawa, T., Nakamura, A. and Yoshikawa, M. (2004). "Antiallergic principles from Thai zedoary: structural requirements of curcuminoids for inhibition of degranulation and effect on the release of TNF-alpha and IL-4 in RBL-2H3 cells." *Bioorg. Med. Chem.* 12, 5891–5898.

55. Foley, S., Navaratnam, S., McGarvey, D.J., Land, E.J., Truscotti, G. and Rice-Evans, C.A. (1999). "Singlet oxygen quenching and the redox properties of hydroxycinnamic acids." *Free Radic. Biol. Med.* 26, 1202–1208.

56. Mori, H., Tanaka, T., Shima, H., Kuniyasu, T. and Takahashi, M. (1986). "Inhibition effect of chlorogenic acid on methylazoxymethanol acetate-induced carcinogenesis in large intestine and liver of hamsters." *Cancer Lett.* 30, 49–54.

INDEX

K

L

M

N

O

P

Q

R